BIG BRAIN BOOK

HOW IT WORKS AND ALL ITS QUIRKS

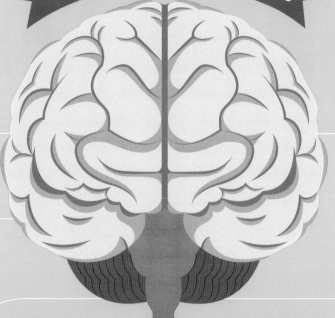

BY LEANNE BOUCHER GILL, PhD

Magination Press · Washington, DC
American Psychological Association

WELCOME TO THE BIG BRAIN BOOK

...written for kids just like you who are interested in big brains! Maybe you have wondered "Why does my heart race when I'm scared?" or "Why do we need sleep?" This book has the answers to these and many other interesting psychology and neuroscience questions about the brain and human behavior. On the next page you will find a list of these questions and the pages that cover the topic. You can read these in order, or you can pick and choose where to start depending on your curiosity! By the end you will be in awe of that squishy organ between your ears!

Each section explains the reasons behind our thoughts, feelings, actions, and senses— your real-life experience explained with the help of psychology and neuroscience. There are graphics, factoids, infographics, and bolded terms that may be new to you with a glossary in the back to reference.

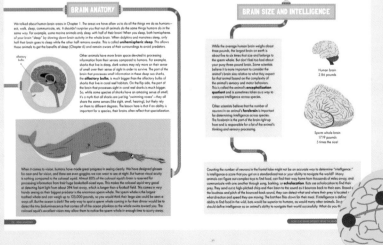

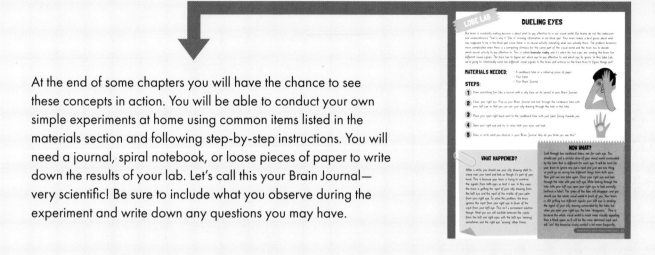

At the end of some chapters you will have the chance to see these concepts in action. You will be able to conduct your own simple experiments at home using common items listed in the materials section and following step-by-step instructions. You will need a journal, spiral notebook, or loose pieces of paper to write down the results of your lab. Let's call this your Brain Journal— very scientific! Be sure to include what you observe during the experiment and write down any questions you may have.

You don't want to miss out on the Dr. Brain sections! Here you'll have the opportunity to meet real scientists who are out there researching these same topics. This is your chance to learn more about the people behind the brain research, what they study, and what they have discovered.

Throughout the book, you can Pick Your Brain with fun facts, because let's face it—the brain is pretty cool! These sections are filled with follow-up questions or facts on a topic related to the chapter. These are things that will really make you ponder next time you think about why your heart races when you are scared or why you need sleep.

LET'S GET STARTED!

This book is dedicated to my parents,
Ronald and Rosemary Boucher,
who always believed in me and my brain—LBG

**Books for Kids From the
American Psychological Association**

Magination Press is a registered trademark of the American Psychological Association.

Order books at maginationpress.org, or call 1-800-374-2721.

Book design by Collaborate Agency Ltd.

Printed by Worzalla, Stevens Point, WI

Library of Congress Cataloging-in-Publication Data

Names: Boucher Gill, Leanne, author.

Title: Big brain book : how it works and all its quirks / Leanne Boucher Gill, PhD.

Description: Washington, DC : Magination Press, 2021. | Summary: "Answers to several common and interesting questions that kids have about the brain and human behavior"-- Provided by publisher.

Identifiers: LCCN 2020033724 (print) | LCCN 2020033725 (ebook) | ISBN 9781433830457 (hardcover) | ISBN 9781433835780 (ebook)

Subjects: LCSH: Brain--Physiology--Juvenile literature. | Brain--Miscellanea--Juvenile literature.

Classification: LCC QP376 .B6914 2021 (print) | LCC QP376 (ebook) | DDC 612.8/2--dc23

LC record available at https://lccn.loc.gov/2020033724

LC ebook record available at https://lccn.loc.gov/2020033725

Manufactured in the United States of America

10 9 8 7 6 5 4 3 2 1

BRAIN ANATOMY

BRAIN AND BODY

WHEN THINGS GO WRONG

CHAPTER 1

WHAT IS

THE BRAIN MADE OF

AND HOW DOES IT WORK?

YOUR BRAIN IS A BEAUTIFUL THING! YOU NEED IT TO...

DREAM

REMEMBER

PLAN

SING

SMELL

YOUR BRAIN IS AMAZING!

Your brain is not a squishy clump of gray matter stuck sloshing around in your skull between your ears. Instead, think of it as an extraordinary little biological computer that controls everything you do and are. The **brain** is an organ that serves as the center of the nervous system in all vertebrate and most invertebrate animals. Your brain is wonderfully complex and is organized into different parts, each with different functions.

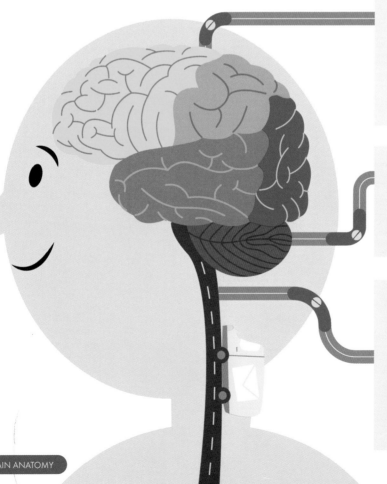

The part that does your remembering, problem solving, thinking, and feeling is called the **cerebrum**, and it fills up most of your skull. The cerebrum is made up of four **lobes**: frontal, parietal, occipital, and temporal.

The **cerebellum**, which keeps you upright and controls balance and coordination, sits at the back of your head, under the cerebrum.

The **brainstem** sits beneath your cerebrum in front of your cerebellum and serves as a relay station between your brain and your **spinal cord**—picture a long highway down your back, carrying messages between your brain and your body.

When you look closer, past the surface, you can see that the brain is made of two kinds of specialized cells: **neurons** and **glial cells**. Neurons are responsible for every thought, word, and action that we think or do. They work a bit like telephones, allowing neurons that are near or far away to talk to one another.

The longest distance your neurons have to communicate is between your big toe and the bottom of your spinal cord! Every time you flex your toe, your brain talks to the spinal cord to tell your toe to move it. Glial cells do a lot of the hard work around the nervous system. They support neurons by insulating them, cleaning up after them, and bringing the neurons things they need to live, like oxygen and food. Glial cells are sort of like the parents of the nervous system. Have you thanked your glial cells (or parents) lately?

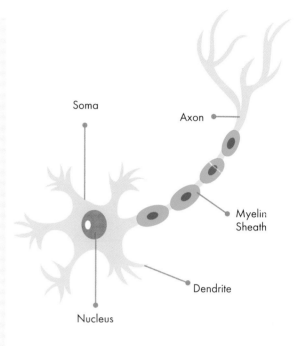

Soma

Axon

Myelin Sheath

Dendrite

Nucleus

NEURONS

synaptic terminal

neurotransmitters

While neurons are shaped differently depending on what they do and where they are located, almost all neurons have similar parts. There are **dendrites**, which are like the long branches of a tree that reach out into the spaces between the neurons and listen for messages being sent their way. When a message is received, the dendrite sends the signal to the neuron's cell body, the **soma**, which decides whether the message was loud enough to send along. From the soma, the signal is sent down the **axon**, which is like a long cable that connects the input side, the dendrites, to the output side, the **synaptic terminal**. The axon is covered in sections by a fatty material called the **myelin sheath**, which prevents the signal from leaking out of the axon. Think of the myelin sheath like an insulated cup that keeps your hot cocoa hot by not allowing the heat to escape. The synaptic terminal then releases chemicals called **neurotransmitters** into the space between the neurons, where they connect with the next neuron's branches, the dendrites. The neurotransmitters are the messengers telling the next neuron that it's their turn to send a message. Then the whole process starts again!

GLIAL CELLS

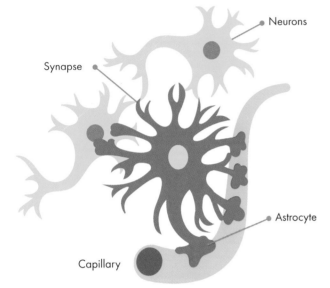

Neurons

Synapse

Astrocyte

Capillary

Glial cells have a huge job! They support the neurons in many ways by holding neurons in place, supplying neurons with oxygen and nutrients, transporting dead neurons away, and providing insulation between neurons. Scientists are discovering that glial cells may also play a role in creative thinking. One type of glial cell, the **astrocyte**, seems to be able to communicate with neurons by releasing a mineral called calcium from what we call their "endfeet." **Astrocytes** have hundreds of endfeet, which they use to communicate with nearby neurons and other astrocytes. Can you picture a tidal wave? Well, imagine a huge wave of calcium leaving the endfeet of an astrocyte and crash-landing on a neuron. Scientists call this a "calcium wave," and they believe that it acts like a messenger, much like the neurotransmitters do for neurons. When the calcium wave meets a neighboring neuron, it leads to the neuron telling other neurons that something is happening, like a thought.

NEURAL COMMUNICATION

KICK!

It may seem weird, but everything we do, from walking to talking to thinking, starts and ends with the release of chemicals, neurotransmitters, that tell neurons to send messages to one another. It's like your brain is the school cafeteria, and all the students are individual neurons talking to one another. If people didn't talk to each other, then things might get confusing. Imagine if you didn't say goodbye to your friend, how would they know you were leaving for class? Or if you didn't tell the lunch worker that you did not want gravy on your mashed potatoes, you might get something you did not like. If a neuron in the part of your brain that controls your foot does not tell your foot to kick, then it won't kick. Everything we do, from kicking to remembering how to kick, is controlled by neurons communicating with one another through signals called **action potentials**.

Action potentials are the currency of the nervous system. Without action potentials, there would be no neural communication. If action potentials are the currency, then the bankers are the neurotransmitters controlling whether they are released by the neuron. Neurotransmitters control whether the action potential message gets passed on to the next neuron, much like when you send a note to your friend in class.

There are certain things you can do that make it more likely that your friend will get your note, like passing the note secretly or when no one is around. On the other hand, if you pass the note to your friend in front of your teacher, it is much less likely that your friend will get to read your note.

In your nervous system, there are two classes of neurotransmitters that make it more or less likely that an action potential will be passed on to the next neuron. **Excitatory neurotransmitters** make it more likely, **inhibitory neurotransmitters** make it less likely. Whenever we do or say or think anything, both of these classes of neurotransmitters are at work encouraging some neurons to send messages and discouraging others.

PICK YOUR BRAIN

Whenever you learn something new, a new connection is formed between two neurons. In fact, as you are reading this book and learning about neurons, you are making new connections in your brain. You can also lose connections between neurons. Have you ever learned something and then forgotten it? If you don't practice skills or actively remember information, you will slowly forget and the connections between the neurons that encoded those skills and information will go away. The brain's ability to change connections between neurons as we experience the world is called **neural plasticity**. We'll learn more about this in Chapter 4 when we cover how we learn.

PICK YOUR BRAIN

The average adult brain weighs just over three pounds and makes up around 2% of one's body weight. When you are born, the brain weighs about four times less, but still makes up about 1% of total body weight. As you grow up, your brain doesn't really get bigger, but many important and tiny changes happen as you learn new skills and facts and have more experiences.

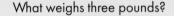
What weighs three pounds?

A six-month old cat

A human brain

A two-slice toaster

A bunch of bananas

A large Jar of pickles

A brick

Some animals have the most unusual brains. There are tiny spiders whose brains are actually so large that they extend out of the head and into their legs! And the sea squirt actually eats part of its own brain once it finds a home in the ocean. More about these strange brains in Chapter 3.

FINAL NOTES FOR YOUR NOGGIN

Your brain is organized into different parts—cerebrum (made up of the frontal, parietal, occipital, and temporal lobes), cerebellum, and brainstem—each with different functions. The brain is made up of specialized cells called neurons and glial cells, which are responsible for all your thoughts and actions. Neurons communicate by making action potentials, which lead to the release of chemicals called neurotransmitters. In future chapters, we will explore how neural communication can be shaped by our experiences and can break down and then build back up after trauma.

MAKE YOUR OWN NEURON

Sometimes it is hard to visualize what neurons look like. We've tried to give you an idea with the pictures in this chapter, but neurons live in the 3D space of your brain, and as we have seen, the connections between them are always changing based on what you've experienced. So, let's make your own neuron—or neurons—to help you learn how these building blocks of the brain really work. Record your observations, questions, and drawings in your Brain Journal.

MATERIALS NEEDED:
- At least 3 pipe cleaners (any color)
- 1 straw
- 6 beads
- Pair of scissors
- Your Brain Journal

STEPS:

1 Take one pipe cleaner and make a circle at one end, leaving the other end long and straight. The circle part is the **soma** and the long straight part is the **axon**.

2 Next, take a second pipe cleaner and cut it into six equal parts. Wrap three pieces around the circle part of the first pipe cleaner. On each of these three pieces, wrap a piece of pipe cleaner around to create a semicircle like pictured to the right. These are the **dendrites**.

3 Cut the straw into small pieces and slide the pieces onto the "axon." The straw bits represent the myelin sheath.

4 Take the third pipe cleaner and fold it in half. Attach it to the end of the "axon" opposite to the "soma" by placing that end of the axon in the middle of the folded pipe cleaner and twisting the folded pipe cleaner around it. These will be your **synaptic terminals**.

5 Finally, thread a few of the beads up each of the "synaptic terminals." These are your **neurotransmitters**.

6 Make a bunch of these and connect them together to mimic neural communication.

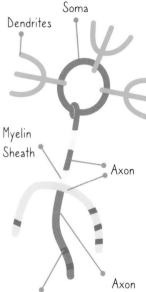

Soma
Dendrites
Myelin Sheath
Axon
Axon
Neurotransmitters

WHAT HAPPENED?

You just made a **motor neuron**, which is just one type of neuron. Your neuron had a soma connected to several branches of dendrites at one end and the synaptic terminals at the other end. The ends were connected by a myelin-covered axon. Many types of neurons have these same parts, but look a little different depending on what the neuron's function is and where it is located.

NOW WHAT?

Make a few of these neurons and put them together to visualize how information goes from one neuron's synaptic terminals to the next neuron's dendrites. Why do you think the motor neuron has this type of structure? Research some other types of neural structures (like **sensory neurons**, pyramidal neurons, olfactory cell neurons, spinal cord neurons, etc.) and try to make these out of the materials listed above. What do you think a neuron's structure has to do with its function?

CHAPTER 2

HOW IS

THE BRAIN

O R G A N I Z E D

?

HOW IS THE BRAIN ORGANIZED?

Your brain is amazingly complex. It allows you to do everything you do—sing, think, act, dance, remember, swim, speak, you name it! Yet, for all its complexity, once you get in there, scientists have discovered that the brain is very well organized. While most areas of the brain are interconnected with one another, there are some distinct areas that are defined by their location. In fact, scientists named some brain areas based on where they are located in the brain; other areas are named after the people who first described them or what object or animal they look like.

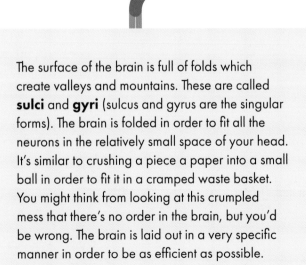

The surface of the brain is full of folds which create valleys and mountains. These are called **sulci** and **gyri** (sulcus and gyrus are the singular forms). The brain is folded in order to fit all the neurons in the relatively small space of your head. It's similar to crushing a piece a paper into a small ball in order to fit it in a cramped waste basket. You might think from looking at this crumpled mess that there's no order in the brain, but you'd be wrong. The brain is laid out in a very specific manner in order to be as efficient as possible.

MAJOR DIVISIONS OF YOUR BRAIN

The brain can be roughly divided into **hindbrain**, **midbrain**, and **forebrain** regions.

At the very back lower part of the brain, rising up from the spinal cord, is the hindbrain. Brain structures located here include the cerebellum, **medulla oblongata**, and the **pons**. These structures help with the basic bodily functions we need to stay alive, including breathing, temperature control, and heart rate. The cerebellum is also involved in other behaviors like balance control and learning. We will talk about how these areas help you learn how to ride a bike in Chapter 9.

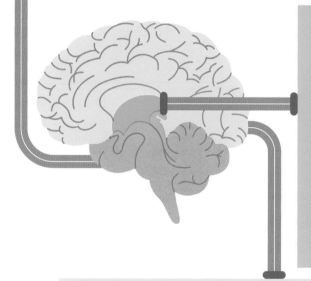

Connecting the hindbrain to the forebrain is the midbrain. The midbrain contains several structures which are important for moving your body and sensing things in the environment. This area is also important in setting your arousal levels. As you go through your day, your arousal level changes based on whether or not you are interested in what you're doing. When your favorite team wins the World Series, your arousal levels go up. When you are bored sitting around the doctor's office waiting to be seen, your arousal levels go down. One time of the day when your arousal levels are way down is when you're sleeping. We'll talk more about that in Chapter 6.

The forebrain is by far the largest part of the brain and contains many structures, including the **thalamus** and **cerebrum**. The thalamus sits just above your midbrain and is responsible for relaying sensory and motor signals. There is a huge crack that runs down the center of your brain from front to back that separates the right and left halves of the brain. The two halves are called **hemispheres**. At the top of the brain, the two hemispheres are not connected. In fact, you could stick your finger between the two halves once you get past the skull and connecting tissue, if that's your thing.

If you kept pushing down, your finger would eventually hit a solid floor. The floor is actually a bundle of nerve fibers called the **corpus callosum** that connects the two hemispheres. Each hemisphere gets sensory information from the opposite side of the body and controls the movement of that side, too. If your right arm got bit by a mosquito, the left hemisphere would know it because your right arm felt it. Seeing as how no one likes to get bit by mosquitos, your right hemisphere would tell your left arm to smash the mosquito on your right arm. The touch information coming from your arm and the "smash the mosquito" motor information to your right arm crisscross in between the spinal cord and medulla.

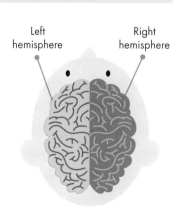

Left hemisphere

Right hemisphere

BRAIN LOBES

The brain is very wrinkly! All the sulci and gyri make it look like a skate park, with some of the dips and transitions being bigger and steeper than others. The **central sulcus** is the longest and deepest dip in the brain, and it is what separates the right and left hemispheres. Other sulci separate other areas of each hemisphere into sections we call lobes. You have four lobes on each side of your brain and each lobe has a special job.

FRONTAL LOBE

At the front of the brain is the **frontal lobe**. This lobe helps you make decisions like whether or not to clean up the mess you made. The frontal lobe also houses your **motor cortex**, the area of the brain that makes you move the vacuum across your rug and throw your dirty clothes in the hamper. As if decision-making and movement weren't enough, this lobe also helps you to remember things, to talk, and to motivate you to keep on going. We'll talk more about these things in future chapters.

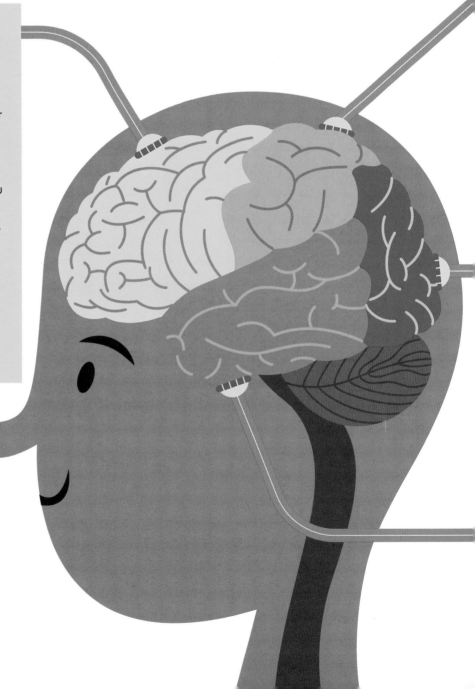

PARIETAL LOBE

Behind the frontal lobe is the **parietal lobe**. This lobe houses your **somatosensory cortex**, the area of the brain where touch information from your body ends up. In fact, there is a map of your body in this area called the **homunculus**, or "little man." The map isn't quite to scale, though, as some body parts take up more space in the somatosensory cortex than others. The hand, for example, takes up much more space in the brain than your back. This leads to you being better able to tell where something is touching you on your hand versus your back, which explains why you can never *quite* reach that itch in the small of your back! We'll get back to this itch in Chapter 11.

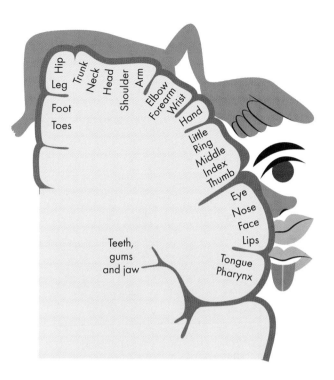

OCCIPITAL LOBE

At the back of your brain is your **occipital lobe**. This area is important for vision. You might think vision is an easy sense to process—you just open your eyes and *voila!* You are seeing! But it takes all kinds of neural processing to watch something like your favorite movie. Some parts of the occipital lobe allow you to see the shiny red and gold color of Ironman's costume, while other parts allow you to watch him zipping across the sky. While most of the time you see things because information from your eyes ends up here, it is not the only way to see. Unfortunately, if you get hit in the head real hard, you might see stars. These stars happen because the neurons in this part of the brain became active due to the trauma of the **concussion**. Your visual neurons don't care how they are activated. If they are activated either by you looking at an object or by physical trauma, you will see something. We'll talk more about concussions in Chapter 22.

TEMPORAL LOBE

Underneath all the other lobes is the **temporal lobe**. This lobe is really important for hearing, learning, and memory. It also houses your **auditory cortex**. There was a man named HM who had surgery to remove both of his temporal lobes in the 1950s. It seems extreme, but HM suffered from **epilepsy**, a brain disorder that causes frequent abnormal brain activity, resulting in uncontrollable seizures; as a result, he couldn't control his body or mind for a period of time. The doctors at the time thought they could cure HM of his epilepsy by taking out his temporal lobes. They were right—the seizures did stop. However, HM was never able to learn any new facts for the rest of his life. We'll learn more about learning in Chapter 4.

PICK YOUR BRAIN

Brain areas get their names based on lots of different reasons. Some parts are named for the objects they look like. The **amygdala** is Greek for "almond" because it looks like the delicious nut. The **hippocampus** is Greek for "seahorse" because it looks like a "sea monster." These two areas are located close to one another in the temporal lobe of your brain. They are part of your **limbic system**, which helps you control your emotions and remember things (see Chapter 9).

FINAL NOTES FOR YOUR NOGGIN

So, you can see that while the brain allows us to do amazingly complex things like sing and dance, it's quite organized. Scientists have divided the brain into regions based on location: the hindbrain, midbrain, and forebrain. The forebrain is the largest and contains the cerebrum, which is home to the four lobes. Within each lobe of the brain are lots of brain structures that have different jobs, from remembering, to talking, to feeling emotions. We'll talk more about what these areas do in future chapters.

LOBE LAB

MAKE A BRAIN

In Chapter 1, we made a neuron. Now it's time to make the whole brain! It is sometimes hard to understand how all the parts of the brain fit together on a 2D piece of paper like in this book. Like our neuron from Chapter 1, it can be useful to make yourself a 3D brain to see how everything works together. That's what we'll do in this lab.

MATERIALS NEEDED:

8 different colors of playdough. It doesn't really matter which structures are which colors, but this gives you an idea:

- Black = spinal cord, medulla oblongata, and pons
- Gray = cerebellum
- White = midbrain
- Orange = corpus callosum

- Red = parietal lobe
- Yellow = temporal lobe
- Green = occipital lobe
- Blue = frontal lobe

- Dental floss
- Your Brain Journal

STEPS:

You are going to make a brain from the bottom up. Try your best to make each brain area look like what you see here in the book. Also, be mindful of the scale of each part (the frontal lobes are the biggest part of brain, for example).
Shape the playdough in the following order:

1 Roll the black playdough into a four-inch tube and make the top part thicker than the bottom. It will look like a snake. The top part is the medulla oblongata and pons and the tail end is the spinal cord.

2 Use gray playdough to make two balls the size of a ping pong ball. These are the cerebellum. Place them on either side of the medulla oblongata (top of the spinal cord you made in step 1).

3 Make an oval the size of an egg with the white playdough. Place it on top of the cerebellum. This is the midbrain.

4 Next, use orange playdough to make a flat pancake the size of a silver-dollar pancake and put that on top of the midbrain, medulla oblongata, and pons. The cerebellum will be sticking out back.

5 Make two of each lobe—one for each hemisphere of the brain. Make sure each half of the brain has a long space between. This is the central sulcus. Put together the lobes in this order:
   a. Parietal lobes on top (red)— a flattened ball of playdough each about the size and shape of a fun-size candy bar.
   b. Temporal lobes on bottom (yellow)— a flattened ball of playdough each about the size and shape of a full-size candy bar.
   c. Occipital lobes in back (green)—each about the size and shape of a small deflated ballon.
   d. Frontal lobes in (you guessed it!) front (blue)—each about the size and shape of a large pancake.
   When you are done, check the brain on page 18. Does your brain look like that?

WHAT HAPPENED?

Congratulations! You just made a brain! You should gain an appreciation of all the different structures and how they work together. The brain should be able to fit in both of your hands, and should be roughly the same size as your brain in your head. Sketch out what the brain looks like in your Brain Journal using your model as a guide. Can you see all the parts you created in your model? What structures are buried deep inside?

NOW WHAT?

Place the brain on a table with the front facing you. Take the dental floss and gently cut through the brain along the central sulcus between the left and right hemispheres, through the corpus callosum, and down into the spinal cord and hindbrain structures. Once you cut the brain in half, you'll be able to see the middle part of the brain, and structures inside. You'll notice that the two hemispheres look alike, but we will learn in the upcoming chapters that some of their functions can be quite different (see Chapter 25). What do you think is the function of the corpus callosum in your model? Why do you think you have two hemispheres with similar brain structures on each side?

CHAPTER 3

HOW

IS MY BRAIN DIFFERENT

FROM THE BRAINS

OF OTHER ANIMALS?

It's your favorite day of the school year—field trip day! This year your class is going to the aquarium. You read all about the special exhibit they have on Large Creatures of the Ocean and can't wait to learn about squid, whales, and sharks. You're also excited to watch the dolphin show and how the corals sway in their tanks. Even though you think it's a weird exhibit to have at an aquarium, you're looking forward to visiting the reptile and bat houses.

You and your friend start to talk about all the ways the animals are different from you both. You wonder if any of the animals have any special abilities, or better abilities than you. If they do, you reason it's because their brains might be set up differently from human brains. You make a mental note to ask someone this very question during the Great Animal Intelligence Show that you'll be watching after eating lunch on the picnic tables near the water.

CENTRAL AND PERIPHERAL NERVOUS SYSTEMS

To compare the brains of animals, we first need to understand the nervous system. The nervous system can be divided into roughly two parts. The **central nervous system** contains the brain and spinal cord. The **peripheral nervous system** includes everything else, including your 12 **cranial nerves** (more on that in Chapter 10), your spinal nerves, and your peripheral nerves. The spinal nerves include both **sensory neurons**, which send information about your senses back to the brain, and **motor neurons**, which send information to your muscles to move. Your peripheral nerves also include sensory and motor neurons and are all over your body, from your feet and legs to your hands and arms to your torso and everything inside. When you have a stomachache, your peripheral sensory neurons in your belly sense you're not feeling well and send a signal up through the spinal nerves to the brain that something is not right. Your brain then sends a signal down through the spinal nerves to your peripheral nerves that make your muscles move to get up and get some medicine or water.

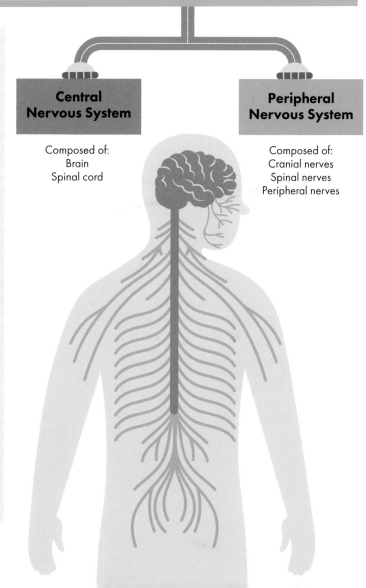

Central Nervous System

Composed of:
Brain
Spinal cord

Peripheral Nervous System

Composed of:
Cranial nerves
Spinal nerves
Peripheral nerves

NERVOUS SYSTEM LOCATIONS

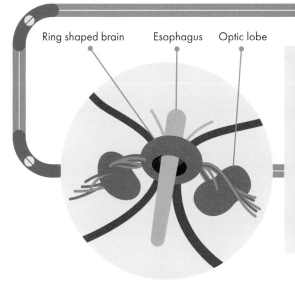

Ring shaped brain   Esophagus   Optic lobe

Most mammals, including humans, and birds have nervous systems set up with the brain located inside the head and the spinal cord in the back that acts as an information highway between the brain and body. In animals other than mammals and birds, nervous systems are a little different. For example, take the giant squid. This cephalopod not only has three hearts, but has a brain shaped like a donut! What is even stranger is that its esophagus goes right through the middle of its donut brain. If the giant squid swallows something really giant, it could potentially damage its brain as the food travels down its esophagus to its stomach.

On land, there are some spiderlings (baby spiders) and very tiny adult spiders whose brains are so large relative to their body size that some of the brains end up in their legs. They're walking brains! On the other hand, cockroaches don't even need a brain to keep walking. If their heads get cut off, the neck seals itself and the cockroach keeps on living, walking around, until it starves to death! Ew!

Back in the ocean, when a sea squirt finally latches onto an object, be it a coral, the ocean floor, or a boat, they eat the part of their nervous system that controls their movement, their **cerebral ganglia.** Ganglia is the scientific term for a bunch of neurons that work together in one area. The sea squirt won't be able to move anymore, but it's a small price to pay for the delicious meal they just ate!

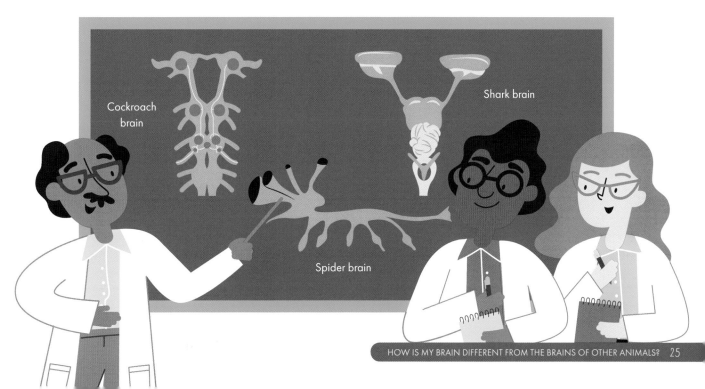

Cockroach brain

Shark brain

Spider brain

We talked about human brain areas in Chapter 1. The areas we have allow us to do all the things we do as humans—eat, walk, sleep, communicate, etc. It shouldn't surprise you that not all animals do the same things humans do in the same way. For example, some marine animals only sleep with half of their brain! When you sleep, both hemispheres of your brain "sleep" by slowing down brain activity in the whole brain. When dolphins and manatees sleep, only half their brain goes to sleep while the other half remains awake. This is called **unihemispheric sleep**. This allows these animals to get the benefits of sleep (Chapter 6) and remain aware of their surroundings to avoid predators.

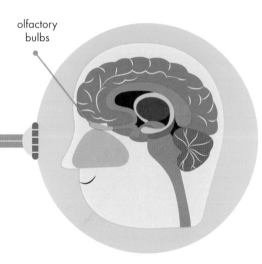

olfactory bulbs

Other animals have more brain space devoted to processing information from their senses compared to humans. For example, sharks that live in deep, dark waters may rely more on their sense of smell over their sense of sight in order to survive. The part of the brain that processes smell information in these deep sea sharks, the **olfactory bulbs**, is much bigger than the olfactory bulbs of sharks that live in coral reef habitats. On the flip side, the part of the brain that processes sight in coral reef sharks is much bigger. So, while some species of sharks have an amazing sense of smell, it's a myth that all sharks are just big "swimming noses"—they all share the same senses (like sight, smell, hearing), but likely rely on them to different degrees. The lesson here is that if an ability is important for a species, their brains often reflect that specialization.

When it comes to vision, humans have made great progress in seeing clearly. We have designed glasses for near and far vision, and there are even goggles we can wear to see at night. But human visual acuity is nothing compared to the colossal squid. Almost 80% of the colossal squid's brain is reserved for processing information from their huge basketball-sized eyes. This makes the colossal squid very good at detecting faint light from about 394 feet away, which is longer than a football field. This comes in very handy seeing as their biggest predator is the enormous sperm whale. The sperm whale is the largest toothed whale and can weigh up to 125,000 pounds, so you would think their large size could be seen a ways off. But the ocean is dark! The only way to spot a sperm whale coming in for their dinner would be to detect the tiny bioluminescence that comes off of the ocean plankton as the whale swims toward you. The colossal squid's excellent vision may allow them to notice the sperm whale in enough time to scurry away.

While the average human brain weighs about three pounds, the largest brain on earth is about five to six times that size and belongs to the sperm whale. But don't feel too bad about your puny three-pound brain. Some scientists believe it is more important to consider the animal's brain size relative to what they expect for that animal based on the complexity of the animal's sensory and motor behaviors. This is called the animal's **encephalization quotient** and is sometimes taken as a way to compare intelligence across species.

Other scientists believe that the number of neurons in an animal's **forebrain** is important for determining intelligence across species. The forebrain is the part of the brain right up front and is responsible for a lot of the animal's thinking and sensory processing.

Human brain
2.86 pounds

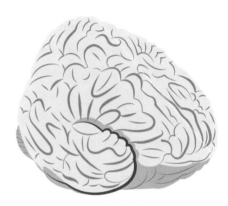

Sperm whale brain
17.19 pounds
5 times the size!

Counting the number of neurons in the frontal lobe might not be an accurate way to determine "intelligence." Is intelligence a score that you get on a standardized test or your ability to navigate the world? Many animals can figure out complex toys to find food, can find their way home from thousands of miles away, and communicate with one another through song, barking, or **echolocation**. Bats use echolocation to find their prey. They send out a high-pitched chirp and then listen to the sound as it bounces back to their ears. Based on the loudness and pitch of the bounced-back sound, they can detect what and where their prey is located and in what direction and speed they are moving. The bat then flies down for their meal. If intelligence is defined as the ability to find food in the wild, bats would be superior to humans, as would many other animals. So perhaps we should define intelligence as an animal's ability to navigate their world successfully. What do you think?

PICK YOUR BRAIN

How Heavy Are Brains?

Species	Average brain weight in pounds	Everyday objects that weigh about the same as the brain
Sperm whale	17	2 gallons of milk
Elephant	11	Large bag of potatoes
Bottle-nose dolphin	3.5	A medium-sized pumpkin
Adult human	3	A 2-slice toaster
Hippopotamus	1.28	¾ of a pineapple
Orangutan	0.82	A little over 3 sticks of butter
Colossal squid	0.22	4 strawberries
Beagle (dog)	0.16	Half a banana
Cat	0.07	7.5 pieces of paper
Great white shark	0.06	$1.25 in quarters
Goldfish	0.0002	A small feather

MEET DR. BRAIN

Dr. Kara Yopak is an ichthyologist, a person who studies fishes. She is interested in how brain size and formation differs between shark species. Her lab has over 1,000 brains from 180 shark, skate, and stingray species that she uses to conduct her comparative neuroanatomy research. She loves what she does and is always excited when she gets a new brain to examine.

Do great white sharks really have a brain that weighs as much as five quarters?

Yes and no. What you have to understand is that there is great variability across species in terms of their overall size and their brain size. The average brain size of an adult great white shark is about 30 grams (0.06 pounds), but it varies greatly depending on body size. The range across different species can be anywhere from .36 grams (0.0008 lbs) in a 540 g (1.19 lbs) cookie cutter shark up to 100+ grams (0.22+ lbs) in a 150 kg (331 lbs) great hammerhead.

What does it mean when we say an animal has a large brain?

Brains require a lot of energy to do their jobs, so if an animal has a large one relative to their body size, that's a big deal. But brain size is not the only thing to consider when comparing across animal species. You really have to look at the relative size of their brain regions and how the brain is organized. I've studied over 180 species of sharks, and I can tell you that the variability in brain organization is tremendous. This is important because you can predict quite a bit about how a shark behaves and what type of environment they live in based on what their brains look like. For example, I can predict differences between shark species just by looking at how much space is taken up by their olfactory bulbs, the part of the brain that processes smell information. Deep sea shark species, like a lanternshark, likely rely less on their visual sense in those deep, dark habitats, than coastal shark species, like a shortfin mako shark. If you look at the brain of a lanternshark, their olfactory bulbs and the parts of the brain that receive input from their electroreceptors are bigger, while in a mako, their visual brain regions are enlarged.

Why study shark brains?

Sharks provide a window into our neural evolutionary past. They were the first group to have the full vertebrate brain "blueprint," which means they have a brain made up of the same fundamental building blocks shared across vertebrates, including you and me! Sharks were also the first vertebrate to develop a cerebellum—a brain region that is involved in motor control and balance. Even more interesting, the architecture of the cerebellum is remarkably similar across many different animals, including birds and mammals. Now you have a fun fact to share at the dinner table tonight—your brain and a shark brain have a lot in common!

Why do you think people are afraid of sharks?

I think people are naturally afraid of what they don't understand or can't see and sharks fit in this category. Most people think of a great white shark when they think of sharks, but they don't realize the huge variability across species. The pygmy shark, for example, is just seven inches long! Sharks also suffer from negative stereotypes, fueled by movies and sensationalist news. All animals have the potential to attack, whether it is your pet dog, a stray cat, or a shark. When an animal is harassed, it is going to defend itself. Sharks do not actively hunt humans. Many sharks' diets consist of smaller fish and plankton—humans are not on their menu!

Where do you get your shark brains?

From wherever we can! In addition to our own fishing, we travel to fishing tournaments and work with local fishermen who catch sharks as by-catch to collect samples. While we're there, we will often do public shark dissections to help bridge the gap between that fear of the unknown and a newfound understanding and appreciation for these animals. We're very passionate about educating the public about the importance of sharks in the marine ecosystem. When we collect the brain for our lab, we will also communicate with other shark researchers who may need other samples, so we can parse out all parts of the shark to various labs for study. No part of the shark ever goes to waste.

Kara Yopak is an assistant professor in the Center for Marine Science at the University of North Carolina Wilmington in Wilmington, NC. Visit her at yopaklab.com

FINAL NOTES FOR YOUR NOGGIN

The human brain is amazing, but so are many other animal brains! Non-human brains may be set up differently than human brains, but we can learn a lot about how different brain anatomies support different sensory and motor abilities of animals. For example, animals that rely on scent to survive have a larger part of their brain that processes smell information, and they are really good at detecting faint scents. While an animal's encephalization quotient or the number of neurons in their forebrain may be related to the intelligence of an animal, it's important to first define what it means to be intelligent. Sure, a bat can use echolocation, but can they play Monopoly?

CHAPTER 4

HOW DO

I LEARN

THINGS?

WILL THIS BE ON THE TEST?

You're sitting in your Civics class listening to your teacher talk about the differences between local, state, and federal governments. You know this is a really important class so that you can learn to be a good citizen (and pass seventh grade), but it is a lot of information! Your teacher has already given you three pages of notes that explain the differences between these governing bodies, and you have no idea how you're going to learn any of it. You decide to ask your friend if she wants to study with you for the upcoming test. She agrees and you meet up at your house the following Saturday. After you get some snacks, you both settle in at the kitchen table surrounded by your books. You open to the chapter on levels of government and get out your notes. You start reading and begin to feel overwhelmed. There's got to be a better way to study than just reading your study materials over and over again. Your friend suggests making notecards and quizzing each other. At first, you don't think this is a good idea since it seems like a lot of extra work, but after a while, you start to actually remember things! You think that your friend might be onto something with her study habits, but why is it easier to remember everything when you study this way compared to just reading your notes over and over again?

CLASSICAL AND OPERANT CONDITIONING (LEARNING)

Think about all the reasons why you do things. Some behaviors seem to be automatic and out of your control, like when you pull your hand away from a hot stove. You have learned that when you touch a hot stove, it hurts! So, you no longer touch hot stoves, or other hot items for that matter. This type of learning that leads to **automatic behaviors** is called **classical conditioning**. Classical conditioning occurs whenever you learn to have an automatic response to something in your environment, like when you pull back your hand from a hot stove, or when your mouth starts to water when you smell a delicious meal, or when your heart starts to race when you get in line for the haunted house at a carnival. It's like your body has learned to anticipate how to react to certain situations.

Other behaviors are more voluntary, like studying for an exam or cleaning your room. This type of learning is called **operant conditioning**. In general, in this type of learning, you do what you do in order to receive a reward or avoid a punishment. If you study and you do well on a test (your reward), then you are more likely to study before your next exam. If you don't clean your room, your parents won't allow you to text with your friends (your punishment), so you are more likely to keep your room clean.

We are evolutionarily programmed to behave in ways that are rewarding—whether it is getting something you like or avoiding something you don't like. Deep inside your brain, you have a "reward system" made up of a bunch of structures, including the **ventral tegmental area** and the **nucleus accumbens**. These areas use the neurotransmitter **dopamine** to signal whether something is rewarding or not. This system floods your brain with dopamine whenever you do something rewarding, whether it is reading a good book, finishing your first 5K, or baking some delicious cookies. This reward system makes sure that behaviors that end in a positive result are more likely to happen.

NEUROBIOLOGY OF LEARNING

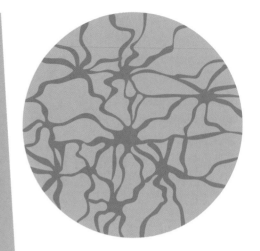

Whenever we are thinking or doing anything, there are groups of neurons that work together to allow you to think your thoughts or do your actions. When multiple groups of neurons are active at the same time, it's because both these groups are needed to complete whatever activity is being performed—whether it is remembering the name of the first president of the United States or tying your shoelaces. When this happens, the connections between the neuron groups are strengthened. So, when the neuron group that encodes the names of the presidents of the United States are active at the same time when you are thinking of George Washington, a connection is formed and you learn something new. Neuroscientists say "neurons that fire together, wire together" to describe this idea.

On the other hand, when neuron groups stop firing at the same time, it's because their joint activity is not needed to do the same task. When that happens, the connection between the two neuron groups is weakened. Have you ever learned something and then forgot it? If you don't practice skills and recall of facts, then you will slowly forget these and the connections between neurons that encoded that information will either go away or be weakened. Neuroscientists say "use it or lose it" to describe this idea.

This fine tuning of connections between neuron groups is what enables us to learn and refine our skills. Think back to when you were learning to ride a bike. It seemed hard to have to remember to pedal with your feet, balance on the seat, steer with your arms, see with your eyes, and try not to crash! But with practice, you got better. This is because the neuron groups you were using to pedal, balance, see, and steer were all working together. Over time, the connections between those neuron groups got stronger. In fact, when you got really good at riding your bike, it's almost as though you didn't have to think about it at all. This behavior is what scientists call **automatic** because it's like your brain is automatically doing something without you thinking about it very much (more on this in Chapter 9).

When you are studying for a test, you are strengthening the connections between neuron groups. It makes sense that the more connections you have, the more you will learn. So when your friend suggested making index cards and testing one another, your brain was actively making more connections between the ideas and facts you were studying and strengthening the existing ones.

One of the best ways to make new strong connections is to think deeply about whatever it is you're studying. That means you can't just read the text. That is quite superficial and will result in little, if any, learning. Instead, try to create connections between the material you're studying and your own interests. With your Civics exam, when you are trying to remember the functions of each branch of the government, think about how your school is a sort of mini-government. Your class is like a local government with the "laws" created and enforced by your teacher. Each class has a different local government. For example, your teacher may allow students to walk around during class, while another teacher may require everyone to stay seated. Your school is like a state government with the "laws" created and enforced by your principal and other administrative staff. For example, the school may decide that all students are required to wear uniforms, but not every school around you has this same rule. Your school district is like the federal government because it creates and enforces "laws" for all schools in the district to follow. For example, your school district decides when the first and last day of schools are for all the schools in your district, but not all schools in your state follow the same calendar. By making the material more personal to you, you are creating stronger connections and will be more likely to remember the material for test time.

PICK YOUR BRAIN

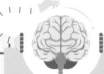

A **mnemonic device** is something people use to remember things. Neuroscience and medical students have to learn the names and order of the 12 cranial nerves (see Chapter 11) as part of their training. A popular mnemonic device for remembering all 12 goes like this: the **O**dor **O**f **O**rangutan **T**errified **T**arzan **A**fter **F**orty **V**oracious **G**orillas **V**iciously **A**ttacked **H**im. The first letter of each word is the same as the first letter of each cranial nerve, in order from one to 12: **O**lfactory, **O**ptic, **O**culomotor, **T**rochlear, **T**rigeminal, **A**bducens, **F**acial, **V**estibulocochlear, **G**lossopharyngeal, **V**agus, **A**ccessory, and **H**ypoglossal.

MEET DR. BRAIN

Dr. Stephen Chew is a cognitive psychologist who studies how people learn. He is the creator of the popular How to Get the Most Out of Studying series on YouTube that students can watch to learn how to study better. His research focuses on the cognitive factors that make students better learners and teachers better teachers.

What are some misconceptions about learning?

There is no hack to studying, there is no easy way to learn, and there is no being lucky on a test. In order for people to learn, they have to practice good study skills. Just because someone is motivated to learn, doesn't mean they will. Developing good study skills takes time and effort. A person can spend a lot of time studying, but if that time is spent on mindless strategies, like reading their notes or textbook over and over again, then they won't learn much.

So what are some good study skills?

Good study skills are a bit counterintuitive. You think that if someone creates flashcards to memorize facts, that they will learn. But the reality is that all they might have done is to memorize a bunch of facts that they will soon forget. In order to learn, one needs to engage in more elaborative study strategies. Concept maps are a great way to study material. A concept map shows you the relationship between facts (see my example on the next page). Seeing the connections between ideas is what leads someone to learning. Another good study strategy is to space out your learning. Don't just sit down and think you're going to study something just one time. If you study some every day, you are much more likely to learn the material than if you study it all at once—even if you end up spending the same amount of time studying. Finally, if

you really want to know if you have learned something, you have to test yourself. You can do this by answering the questions you may find in your textbook, by having someone else quiz you, or even better, write down everything you think you know about the material. Just close your notes and textbook and get out a blank piece of paper and write it all down. This will show you what you really know and what you don't know. Use this feedback honestly so that you can go back and spend time studying the things you weren't sure about.

Do you have any advice for teachers?

There are so many great teaching resources out there. I think giving honest and detailed feedback is incredibly helpful for students to understand what they know and what they don't know. Oftentimes, students have poor metacognition where they don't know that they don't know something. One thing teachers can do to help students with their metacognition is to give students plenty of opportunities to test their knowledge through low-stakes assignments and quizzes. Following this up with feedback and more opportunities for practice will really help students to learn the material well.

Dr. Stephen Chew is a professor of psychology at Samford University in Birmingham, AL. Visit him at samford.edu/how-to-study

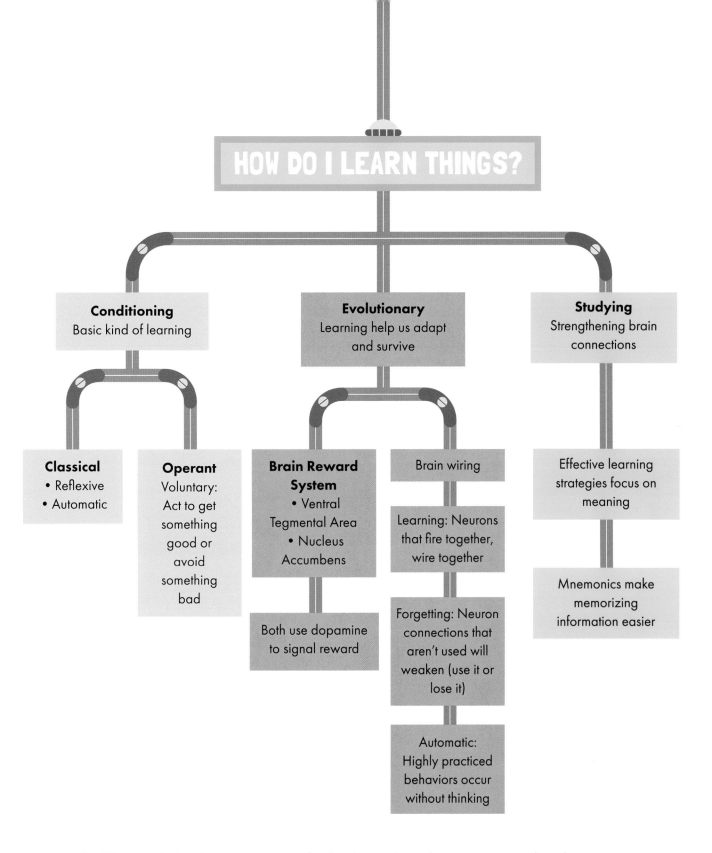

HOW DO I LEARN THINGS?

Conditioning
Basic kind of learning

Classical
• Reflexive
• Automatic

Operant
Voluntary:
Act to get
something
good or
avoid
something
bad

Evolutionary
Learning help us adapt
and survive

**Brain Reward
System**
• Ventral
Tegmental Area
• Nucleus
Accumbens

Both use dopamine
to signal reward

Brain wiring

Learning: Neurons
that fire together,
wire together

Forgetting: Neuron
connections that
aren't used will
weaken (use it or
lose it)

Automatic:
Highly practiced
behaviors occur
without thinking

Studying
Strengthening brain
connections

Effective learning
strategies focus on
meaning

Mnemonics make
memorizing
information easier

Dr. Chew made the above concept map for this chapter. Note that concept maps show the connections between ideas. They can be as detailed as you like, and they don't have to be neat or fancy!

FINAL NOTES FOR YOUR NOGGIN

People learn in a variety of ways. Sometimes the learning is automatic through classical conditioning; other times, learning is based on a reward system through operant conditioning. No matter which way you learn, your brain is constantly making new connections between neuron groups—"neurons that fire together, wire together." If you don't practice what you've learned, those connections will be weakened, and you might forget the material entirely—"use it or lose it". When studying, don't just read your text, but rather think of ways to relate the material back to yourself to ensure stronger connections and increase your learning.

LOBE LAB

REMEMBERING ALL OVER AGAIN

Even when we forget information, it is often not lost forever. Have you ever learned something and then had to relearn it? It usually takes a shorter amount of time to relearn the information. Hermann Ebbinghaus, a German scientist from the 1800s, called this *savings* based on his observation that even when information seems to be forgotten, it takes less time to relearn the information. In this Lobe Lab, we're going to measure savings when trying to remember made up words.

MATERIALS NEEDED:
- 12 index cards (or pieces of paper)
- 12 made up words that follow the pattern consonant-vowel-consonant (i.e. TIF)
- A friend or family member
- Your Brain Journal

STEPS:

1 In your Brain Journal, draw a line down the middle of the page to create two columns. In the first column write down "Trial Number" and in the second column write "Number of Words Remembered".

2 Write down one made up word on each of the 12 index cards.

3 Show your friend one index card at a time.

4 After you have shown your friend all 12 index cards, have them say out loud all the words they can remember from the list. This is Trial #1. Note in your Brain Journal the number of words your friend remembered correctly for this trial.

5 Keep repeating steps 3 and 4 until your friend remembers all 12 words. Each time you repeat this, it's a new trial and you'll write down the number of words your friend remembered.

6 Stop the procedure when your friend gets all 12 correct. How many trials did it take to get to 12 correct words?

7 Wait two days.

8 Retest your friend by repeating steps 3-6. Be sure to record your trial results (number of trials and number of correct words) until your friend gets to 12 correct words.

WHAT HAPPENED?

Your friend should have had an easier time remembering all of the words the second day of the lab. This is because while information may be forgotten, it is not gone forever. This is what Ebbinghaus called savings. This is good news for you, especially when you need to study for an exam where you haven't looked at the material in a while. It will take you a shorter amount of time to relearn the material, leaving you more time to conduct experiments like this!

NOW WHAT?

Try this experiment again but retest your friend after five days. Does the length of time between the two test days make a difference in how much they *saved* in their memory? What if your friend uses a mnemonic device to remember the words? They can make up a song or rhyme, or even come up with a story that includes all 12 nonsense words. It will make for a nonsense story, for sure, but it will lead to a memorable one!

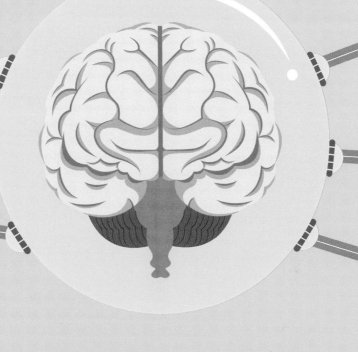

CHAPTER 5

HOW DO SCIENTISTS

MEASURE

BRAIN ACTIVITY?

HERE'S LOOKING AT YOU, BRAIN!

Have you ever wondered what your brain looks like or how it works? You probably have a good idea about its size because you know how big your head is, but you can't tell what the brain looks like or what it's doing just by looking at your head. There are all sorts of methods that scientists use to study the brain. Brain scans are used when a scientist wants to study what is happening inside a human brain. **Structural brain scans** allow people to see the structures of the brain, similar to a picture. **Functional brain scans** allow people to see how the brain responds over time, similar to a movie. Scientists can also measure brain activity using **electrodes**, tiny wires that can record the electrical activity of the brain. The electrodes are either placed on the scalp or directly into the brain.

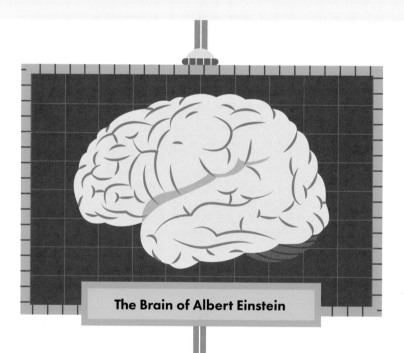

The Brain of Albert Einstein

Scientists often rely on many different ways of measuring brain activity to understand how the brain works. For example, scientists were curious about what made Albert Einstein so good at math, and so they compared the structure of his brain to other people's brains. They found that Einstein's temporal lobe was wider than control brains by about 1 cm—the width of your fingernail! We know from functional scans and electrode recordings that the temporal lobe is important for memory and numerical thinking. Some scientists believe Einstein was good at math partially because his temporal lobe was bigger than average by a fingernail!

BRAIN SCANS

So, you've got a brain and you want to study it. Many of the ways that scientists study the brain are the same ways that doctors look at other parts of your body, like when you sprain an ankle, break an arm, or have unexplained belly aches. Your brain is made up of neural tissue and fluid. In general, different types of tissue and fluid show up differently in each type of scan due to differences in the density and electromagnetic properties of the brain tissue and fluid. Depending on the type of scan, one can see individual structures of the brain, the presence of a tumor or fluid, and how the brain tissue responds when a person is doing something like listening to music, watching a movie, or remembering just how good the mac-n-cheese was at dinner last night.

CT SCAN

CT stands for Computerized Tomography. When a person has a **CT scan** of their brain, they are asked to lie down on a table as still as possible. The table then slowly moves into a machine that resembles a large donut. When the CT scan starts, the machine makes a lot of noise as it takes X-rays from a bunch of different angles. A computer program then combines the images into one 3D image. A CT scan is good enough to detect the presence of a tumor or fluid in a part of the brain where there shouldn't be any.

MRI SCAN

MRI stands for Magnetic Resonance Imaging. Similar to a CT, a person lays down on a table and enters into a machine that looks like a big donut. Instead of X-rays, however, the MRI uses a strong magnet to align the protons in your brain. Protons are tiny particles that are inside every cell in your body. The machine sends out a quick and painless electromagnetic pulse that puts the protons out of whack. An MRI measures the time and energy it takes for the protons to realign themselves. An MRI takes a really clear picture of the brain—you can see individual brain structures deep inside the brain and the gyri and sulci on the surface of the brain.

FMRI SCAN

fMRI stands for Functional Magnetic Resonance Imaging. A person getting an fMRI scan undergoes the same procedure as someone getting an MRI. However, the difference is that the fMRI takes a picture of the brain *over time*, so it's like you get to see a movie of what the brain is doing when you're thinking about that delicious mac-n-cheese dinner.

ELECTRODE RECORDINGS

Pictures and movies of the brain are cool, but when a scientist needs to measure what individual neurons are doing, electrode recordings are used. When neurons generate action potentials, electricity is generated as well. Electrodes measure the electrical activity of neurons either through the scalp or when placed directly into the brain tissue.

EEG

EEG stands for electroencephalogram. When a person gets an EEG, small metal discs are attached to electrodes placed directly onto the scalp. The electrodes then record the brain activity over time to a computer. Doctors can use an EEG to measure if your brain activity is abnormal, like it may be if you are having a seizure. Seizures are abnormal bursts of electrical activity and can cause brain damage if left untreated. Scientists use an EEG to measure how the brain reacts under certain conditions. For example, if your best friend came up behind you and yelled, "BOO!", the parts of your brain that make you scared would be highly active, as might the parts of your brain that allow you to yell at your so-called friend!

Scientists can measure the electrical activity of a single neuron ("unit") by lowering an electrode directly into your brain. As the electrode moves through the brain, it records the electrical activity that it "hears." Your brain is always making noise because there is always electrical activity happening in your brain, even when you're sleeping or bored. When an electrode gets close to a neuron, it can record its action potential. Action potentials are fast! From start to finish, they take one millisecond! A housefly takes three times as long to flap its wings. Scientists can measure the activity of a neuron to figure out what it is communicating. Sensory neurons communicate that a sensory stimulus, like a light or a sound, happened. Motor neurons communicate that it's time to move a body part, like a leg or an arm. Other neurons communicate that you're getting hungry or that you forgot to do your homework.

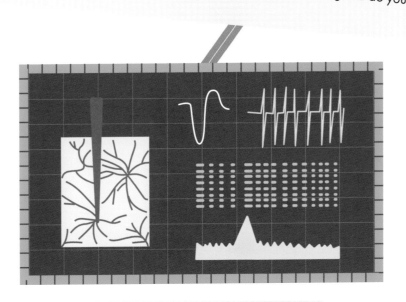

PICK YOUR BRAIN

When you look at a picture of a brain, you need to know from which direction the brain is being imaged. A **horizontal image** shows the brain from top to bottom. A **coronal image** shows the brain from front to back. A **sagittal image** shows the brain from side to side.

horizontal   coronal   sagittal

MEET DR. BRAIN

Dr. Erik Emeric is a research specialist at Johns Hopkins University in Maryland. Scientists like Dr. Erik Emeric use single-unit recording to measure brain activity while monkeys make decisions. Whenever you have a choice between two options, you have a decision to make. Imagine if your mom said you could either have one candy bar now or you could wait an hour and then have two candy bars. Would you choose the smaller reward now or wait for a larger reward later on? These are the types of questions that Dr. Emeric seeks to answer in his research. How easy is it to wait for a bigger reward and which neurons control that decision?

Where in the brain do you record from?
We record from areas of the brain that are involved in motor decision-making. These areas are in the front part of the brain and include the motor cortex, basal ganglia, and the supplementary motor area.

What are you hoping to discover?
I study how these motor neurons are involved in self-control and temptation. We give the animal a choice to either get a small amount of juice now or to get a large amount of juice after waiting a bit. The best choice is to wait, but waiting can be difficult. We want to know which neurons are involved in the decision to give into temptation and which are controlling the decision to wait for the bigger reward.

How do you find a neuron to record from?
We slowly lower an electrode into the brain which records the electrical signals coming from neurons.

We amplify the signal through a speaker and when a neuron produces an action potential, you hear a "pop!" It's a little like listening to rice krispies crackle away in milk.

What does it feel like when you find a neuron?
Since it takes so long to find a neuron, I first get excited and then worried. I'm worried that I might lose the signal because of the tiny movements the brain makes every time blood is pumped into the brain tissue. If I don't have the neuron for a good amount of time, then I won't be able to record enough trials so I can make sense of the data. You need a lot of trials to figure out how a single neuron is involved in whatever behavior you are studying.

How long can you record from a neuron?
My personal best was about 30 minutes, but my personal worst was five seconds! You just never know!

FINAL NOTES FOR YOUR NOGGIN

There are many ways that scientists and doctors can measure brain activity. Brain scans allow us to get a good picture or video of the whole brain. Electrode recordings allow us to measure brain activity directly through the scalp or in the brain. Many of these techniques are relatively new, only being discovered within the last half century, and new techniques are being discovered all the time. In future chapters, we'll rely on these brain measures to learn about how and why the brain works the way it does.

DIY BRAIN SCANNING INVESTIGATION

Brain scanning is not something you can try on your own, unfortunately. You need a lot of expensive equipment and specialized training to measure brain activity. Luckily, we have the internet! In this Lobe Lab, you're going to search the internet for brain images to see how they *measure up* to each other.

MATERIALS NEEDED:

- Computer or mobile device with internet access
- Your Brain Journal

STEPS:

1 Open your favorite internet search browser and search for the following:
- Head CT scan
- Head MRI scan
- Head fMRI scan

Some great websites to explore include:
- https://kidshealth.org/en/kids/med-videos-landing.html
- https://faculty.washington.edu/chudler/image.html

2 Note in your Brain Journal the type of scan, the orientation of the scan, and your observations by creating a table with three columns labelled "Type of Scan," "Orientation of Scan," and "Observations." Then write down your impressions of the scan—what does it look like? Is it a clear picture? What brain structures can you see?

3 Compare multiple fMRI brain scans and note what the person was doing while their brains were being scanned. You'll see splotches of color on the fMRI brain scans which represent the level of brain activity in a particular brain area. Generally, brighter colors mean the neurons in that area of the brain are more active.

WHAT HAPPENED?

You should have noticed that different types of brain scans gave clearer pictures than others. Which scan gave you the clearest ones? Those are the scans that take pictures of the brain at a higher resolution, sort of like when you watch your favorite TV show in high definition compared to standard definition. In the fMRI scans, did you notice if there was a relationship between what the person was doing while getting their brain scanned and which areas were more active?

NOW WHAT?

The more brains you look at, the better you'll become at recognizing different brain features. Keep on searching for brain scans, adding terms like "tumor" or "Alzheimer's disease" or "Einstein" to your search. Do you notice if there are particular brain areas that keep coming up if you search for things like "working memory" or "visual imagery"? Add a brain structure, such as the hippocampus or corpus callosum, to your internet search. Note how the structure is easier to see if the scan is taken from a horizontal, sagittal, or coronal view.

CHAPTER 6

WHY DO

WE NEED TO

SLEEP?

I DON'T WANT TO GO TO BED!

You're playing a wicked game of chess in the game room with your best friend when all of a sudden, you hear the dreaded words from your father, "Time for bed, you two!" You yell back "But Dad, we're almost done!" He then tells you that you have been saying that for the last half hour and it's time *now* to go to bed or your friend won't be able to sleep over again. You and your friend sigh and trek up the stairs to your room. You shut out the lights, but you both know you won't be sleeping any time soon! You're having a sleepover, after all. Why waste all that time sleeping? You are woken up the next day by your dad yelling, "Pancakes! Rise and shine!" You're super tired. You and your friend stayed up late talking. You get up and bump your head on an open drawer and the pain is searing! All you want to do is go back to sleep. Why does sleep seem so great in the morning, but not so great at night?

SLEEPING

It may seem like all you're doing when you're sleeping is just lying there, but when you're sleeping your brain and body are undergoing all sorts of changes. Scientists talk about **stages of sleep** and define these stages based on brain and muscle activity, which is measured through **polysomnography** (say that five times fast!). Polysomnography measures heart rate, breathing, brain activity, and leg and eye movements. All this information helps researchers determine which stage of sleep a person is in and how long they are in each stage.

Adult Brain Waves

∿∿∿∿∿∿∿∿∿∿	Awake
∿∿∿∿∿∿∿∿∿∿	Drowsy
∿∿∿∿∿∿∿∿∿∿	Stage 1 sleep
∿∿∿∿∿∿∿∿∿∿	Stage 2 sleep
∿∿∿∿∿∿∿∿∿∿	Stage 3 and Stage 4 sleep (slow-wave sleep)
∿∿∿∿∿∿∿∿∿∿	Stage 5 (REM sleep)

As you start to nod off to sleep, you begin to lose some of your muscle tone. If you're at school, you can see when someone is dozing off into Stage 1 when their head starts to bob in class. Their neck muscles begin to give out and they can no longer hold their head up. Stage 2 is a deeper stage of sleep and is marked by a slowing in brain activity that is measured using an EEG machine (see Chapter 5). If you call to your friend who is sleeping in this stage, you can still wake them up, but you'll need to yell pretty loud because their senses are starting to dull.

Stages 3 and 4 are both deep sleep stages. It's very difficult to wake someone up in this stage. This is also the stage in which people might sleepwalk **(somnambulism)** or sleep-talk **(somniloquy)**. Both of these things are perfectly normal, and it is not true that if you wake someone up when they are sleepwalking, they will die. They will be confused, but it's hardly fatal! If you have ever experienced a nightmare or **night terror**, it most likely occurred in this stage. Night terrors are extremely frightening nightmares. Night terrors are relatively rare and usually happen between the ages of four and 12. Scientists think they happen because our nervous system is over-active. The good news is that most kids don't remember the night terror in the morning and that they stop having night terrors as they grow older and their nervous systems mature.

Stage 5 is the last stage of sleep and is sometimes called **paradoxical sleep** because your brain activity looks exactly like it does when you're awake! So you can't tell whether a person is awake or asleep just by looking at the brain activity on the EEG. Instead, you have to look at the muscle activity and what you'll see is that the muscle activity in your arms and legs is completely silent! People in Stage 5 sleep don't toss and turn or move their arms about because they are paralyzed. The only muscle group that is working is the one that controls eye movements. If you look at your dad's eyes while he's "resting" on the couch and he's in Stage 5, you'll see his eyes moving like crazy under his eyelids. This is why this stage is also called the **Rapid Eye Movement (REM)** stage.

WHY SLEEP?

It seems like such a chore having to go to bed. You have to brush your teeth and change into your pjs just so you can waste a bunch of time lying down with your eyes closed not texting your friends or watching your favorite tv show. Well, even though it may seem as though you're doing nothing, your body is hard at work while you're sleeping! Scientists believe there are many reasons for why we sleep and one of them has to do with the fact that your body needs time to recoup, reorganize, and get ready for another day of active thinking and moving. While you're sleeping, your body is growing muscles, repairing tissues, and restoring hormone levels. Another very important thing that happens while we're sleeping is that your brain is solidifying all the information you learned during the day into your memory. Lots of experiments show that people are better able to remember things after a good night's sleep. Scientists think this is because the connections between neurons in your brain are strengthened while you sleep (see Chapter 6). This is why pulling an all-nighter before an exam is a bad idea. Without sleep, your brain can't encode all the information that you're studying into memory. You're better off getting sleep between studying and taking your trigonometry test!

SLEEP DEPRIVATION

Many people have trouble sleeping. Some have problems falling asleep and others have problems staying asleep. Both of these issues can lead to **sleep deprivation**. There are many negative consequences of sleep deprivation, including feeling overly emotional, being more sensitive to pain, hallucinating, and having poor concentration and memory. Randy Gardner was a high school student and in 1964, he remained awake for 11 days and 25 minutes (264.4 hours). Randy's attempt was closely watched by researchers who documented his physical and mental well-being throughout his entire time awake. As time went on, Randy became moody, paranoid, and started to hallucinate. He also was unable to concentrate, and his memory was poor. On the 11th day, he was asked to count backwards from 100 by 7s: 100...93...86...79...72...65. He just stopped at 65 and was silent. When asked why he stopped, he said he couldn't remember what he was doing! His memory and concentration had gotten so bad that he couldn't go on.

People have tried to break Randy Gardner's sleep deprivation record, and some claim to have succeeded. Maureen Weston was able to stay awake for 449 hours in a rocking chair marathon in 1977 and holds the *Guinness World Record* for staying awake the longest. However, due to the negative consequences of not sleeping for long periods of time, the *Guinness Book of World Records* no longer records how long people are able to stay awake. It's also hard to verify that people are actually awake for long periods of time since people can slip into brief **microsleeps** where the person doesn't even realize they fell asleep!

PICK YOUR BRAIN

You will be asleep for about a third of your entire life! That's like:

- 1 side of a triangle
- 1 foot in a yard
- 1 of the three little piggies
- 1 goal in a hat trick
- 1 hole in a bowling ball
- 1 Musketeer

1 foot

1 yard

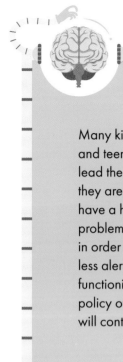

PICK YOUR BRAIN

Many kids are sleep deprived and it's not their fault. Pre-teens and teens experience changes in their sleep needs which lead them to experience a sleep phase delay. This means they are naturally more alert later at night and sometimes have a hard time falling asleep before 11 pm. This is a problem only because these same teens need to wake early in order to go to school in the morning. Not only are teens less alert first thing in the morning, but they are probably functioning on little sleep. Until school districts change the policy of an early morning start time, sleep deprived students will continue to roam the halls of our schools.

RING!!!

MEET DR. BRAIN

Dr. Jaime Tartar is a neuroscientist who studies how and why people sleep. Do you ever notice how you don't feel good after you have had a bad night's sleep? Or that if you're sick, all you want to do is sleep? These are the types of questions that Dr. Tartar studies by researching how our sleeping behavior can impact our physical and mental health.

Why did you decide to study sleeping behavior?

I had trouble sleeping in college and I wanted to know why.

What is a typical sleep experiment like?

We get people to wear special watches that measure different aspects of sleep behavior. The watches measure things like how long they sleep, how long it takes them to fall asleep, and how often they wake up after they fall asleep. They then come into my lab and we collect saliva samples by collecting their spit into a small tube. Later on, we analyze the saliva samples using a special lab technique that measures their immune system functioning and their level of stress hormones. We find that people who are sleep deprived get sick more often and are moody.

How does sleep affect your health?

Sleeping is your brain's way of taking out the trash. While you're awake, the levels of toxic proteins in your brain increase, and as you sleep these levels decrease. If you don't get a lot of sleep, then you might have too many toxic proteins in your brain which can lead to changes in the way your cells work (known as gene expression). These changes in the way cells work can lead to both physical and mental illnesses.

Do you have any advice for people who have a hard time going to sleep?

That's the million dollar question! The best advice is to sleep in a cool, dark room. Also, don't drink any caffeine after noon and try to keep the lights down low at night. Finally, most people can't fall asleep because they are anxious about something. Try to clear your mind before bed by doing some yoga or meditation.

Dr. Jaime Tartar is a professor of psychology and neuroscience at Nova Southeastern University in Fort Lauderdale, Florida.

FINAL NOTES FOR YOUR NOGGIN

Humans spend about a third of our lives sleeping. Even though it seems like a waste of time, your body and brain do a lot during your sleeping hours. Muscles and tissues are created and repaired, hormone levels are restored, and connections between neurons are formed. All of this leads to healthier bodies and better memory and focus during the day. You go through five stages of sleep in which your brain and muscle activity changes. Sleep deprivation can lead to cognitive and mental health changes, including poor memory and attention, hallucinations, and an increase in pain sensitivity. So, if you have to hit your head on that open drawer, make sure you get a good night's sleep first!

CHAPTER 7

HOW CAN WE SEE IN

3D

AT THE MOVIES?

LET'S GO TO THE MOVIES!

Don't you just love going to the movies? There's popcorn, candy, a big screen, loud sound effects, and when you're lucky, the movie is in 3D! The way the scenes pop out of the screen makes you feel like you're right there in the room with the actors. Even the silly 3D special effects, like when something is thrown right at you or when you are swerving in the car in a high-speed car chase, are exhilarating. The amazing fact is that this 3D experience is something that is created in your brain! When your eyes look at the screen, each sees something like a painting in 2D. Your brain then does the hard work of putting the two 2D paintings together to make everything look like it has depth. This happens not only when you're at a 3D movie, but whenever you open your eyes and see that your mom is standing three feet in front of you giving you the stink eye because you've been playing video games all day long instead of doing your homework!

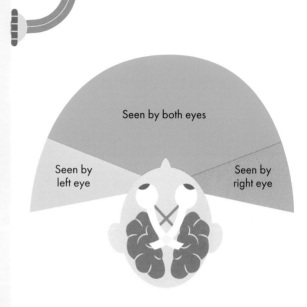

WHAT IS DEPTH PERCEPTION?

If you play nearly any sport, you know how important depth perception is to hit the ball or score that goal. Good depth perception allows you to judge how far away something is and how fast it is moving towards you, so you can reach up and hit the volleyball over the net. You even have depth perception in the water—swimmers can judge how far away the edge of the pool is before they have to flip over and start swimming the other way.

There are all sorts of ways for how our brain figures out if something is near or far. Some ways only require one eye and other ways require both eyes. Things that are farther away look smaller and are fuzzier. With one eye, your brain compares the sizes and clarity of objects and makes a rough guess of how far away the object is from you. If you have two baseballs that are at different distances from you, the closer one will appear larger and you will be able to see the red stitches clearly. When the baseball is farther away, it looks smaller and the red stitches don't look as clear. You can see quite a bit in depth with just one eye—ask any pirate! But excellent depth perception happens when you have two eyes working together. This is why outfielders should always keep both eyes open when they catch that foul ball.

Did you know that you are able to see in 3D because of where your eyes are on your head? The placement of the eyes is important because it determines the size of your **visual field**, or how much you can see without moving your head. Our visual fields go from left to right about 190 degrees. Think of your favorite pizza. If you ate the whole thing, you'd eat 360 degrees of pizza. The size of your visual field is just over half of the pizza. Not a whole pizza, but still enough to make you full. Animals that have eyes on the sides of their heads, like horses and rabbits, have a much bigger piece of the pie. In fact, rabbits have almost the whole 360 degrees of the pizza! They only have blind spots just in front of them and just behind them. That's why they are so hard to catch—they can see you coming from almost any angle.

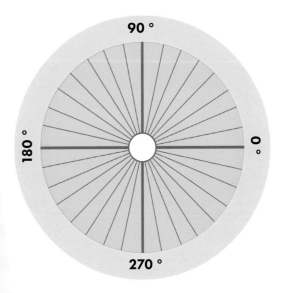

Since you have two eyes in the front of your head, there is a large part of the visual field that both of the eyes see. This **binocular field of vision** is about the size of two slices of pizza. The remaining slices are only seen by one eye, your **monocular field of vision**. We need our two eyes to see in depth. So, animals with large fields of binocular vision have excellent depth perception. Humans and other animals with forward facing eyes, like cats and dogs, have a large portion of their visual field that is binocular. This explains why your dog is able to catch the ball most of the time when you play catch. In contrast, while animals with large visual fields, like rabbits and horses, can see almost all around them, most of their visual field is monocular, so they have poor depth perception. That is why you should never play catch with a rabbit!

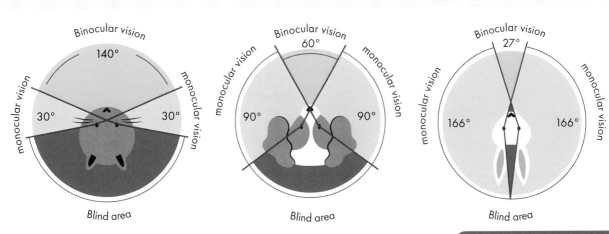

You might think that it's a bit redundant for both eyes to see the same parts of the world, but while your eyes may see the same things, the position of those things that your eyes see is slightly different. You can convince yourself that your two eyes see slightly offset images of the world by doing the following exercise. Extend your arm out in front of yourself and give yourself a thumbs up. Close one eye and stare at your thumb. Line up your thumb with an object in the background—a wall, a water bottle, the TV, whatever. Without taking your eye off your thumb, notice how the position of the objects behind your thumb jump back and forth as you open and close each eye in turn. You can see how the images are just slightly off from one another. Yet, when you open both eyes, you mostly just see one version of the world.

When your brain gets the two images of the world from your eyes, it compares the two images and attempts to line up bits of the images together. For example, when you were giving yourself a thumbs up, your brain lined up the sides of your thumb that your left and right eye were seeing so you saw one thumb in front of you. Being able to see in depth because your brain gets two slightly offset images is called **stereopsis**. However, when the brain tried to match up the sides of the objects behind your thumb, it was unable to do so, and you saw two of those objects. When your brain can't find matches in the left and right eye images and you see two objects when there is only one, it's called double vision, or **diplopia**. In extreme cases, like when the images of the left and right eye are completely different, your brain will actually ignore one of your eyes. We will talk more about this in Chapter 12 when we discuss what it means to be conscious.

So, how do the movie studios take advantage of your binocular vision to give you the 3D experience we have come to know and love at the movie theater? The different types of 3D movie technologies, including IMAX and RealD 3D, use different hardware in terms of the types of glasses you wear, the cameras that record the film, and the screen itself. However, all 3D technologies take advantage of our brain's ability to put together the two images taken in by both eyes.

You might not know it, but the two lenses on your 3D glasses let in different images from the movie. Have you ever looked at the screen without your 3D glasses on? It looks fuzzy because there are actually two different versions of the movie on the screen. The 3D glasses make sure that your left eye sees one image and your right eye sees the other. Your brain then compares the images and voila! You see the movie in depth!

PICK YOUR BRAIN

Look at the picture. It looks as though those people are different heights, but they are actually the same height! It's a visual illusion. Your brain thought the person on the right was a lot farther away because the lines give the illusion of depth. Since your brain thinks the person on the right is farther away, it assumes the person must also be a lot bigger. To convince yourself that the people are the same height, measure them!

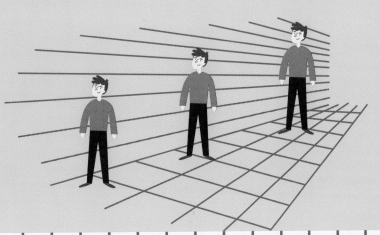

FINAL NOTES FOR YOUR NOGGIN

In order to see a 3D movie, your brain needs to get two different pictures from your left and your right eyes. The two pictures are pretty much the same, except that the positions of the things you are seeing are slightly off from one another. The 3D glasses make sure that your two eyes see these two different versions of the movie. Your brain matches up the two pictures, and you see and feel as though you're riding the roller coaster around the loop de loop!

FIELD OF VISION

In this lab, we're going to measure the size of your visual field. You don't need any fancy medical equipment to do this, just your eyes, your thumb, some object, and a willingness to look funny!

MATERIALS NEEDED:
- Your eyes
- Your thumb
- A small object (it could be a water bottle, a pencil, a stuffed animal, whatever)
- Your Brain Journal

STEPS:

1 In your journal, draw a circle. Draw lines in the circle that divide it into four quadrants. One line should go from the bottom to the top and the other line should go from left to right. Now you should write some of the degrees around the circle, much like you see here. At the top of the vertical line, write zero and at the bottom of it, write 180. On the right side of the horizontal line, write 90 and on the left side, write 270. You can fill in some of the degrees in between these if you want.

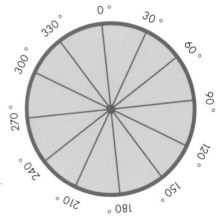

2 Focus your eyes on an object in front of you and keep them still.

3 Close your left eye and put your right arm out in front of you so that your thumb is covering the object. This is where 0 degrees is on your circle.

4 Move your right thumb slowly from the object to the right until you can no longer see your thumb. That is the end of your right eye's visual field. Record on the circle the degree at which you think your right eye's visual field ends.

5 Put your right thumb all the way at the right end of your visual field and slowly move it back toward the object at the center and keep moving it past the object until you can no longer see your right thumb. That is the other end of your right eye's visual field. Now record that on the circle. You have just mapped your right monocular visual field.

6 Now let's do the same for your left eye. Close your right eye and put your left arm out in front of you so that your thumb is covering the object once again at 0 degrees.

7 Move your left thumb slowly from the object to the left until you can no longer see your thumb. That is the end of your left eye's visual field. Record on the circle the degree at which you think your left eye's visual field ends.

8 Now, put left right thumb all the way at the left end of your visual field and slowly move it back toward the object at the center and keep moving it past the object until you can no longer see your left thumb. That is the other end of your left eye's visual field. Now record that on the circle. You have just mapped your left monocular visual field.

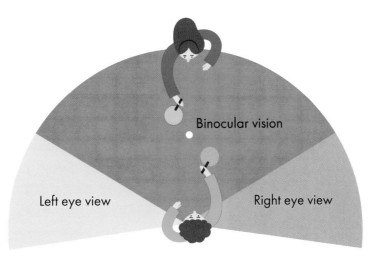

Binocular vision

Left eye view   Right eye view

WHAT HAPPENED?

Look at your map. You should have a large portion in the middle of your visual field that overlaps between your left and right eyes. This is your binocular field of vision and should span about 120 degrees (60 degrees to the right and left of 0 degrees) There should be smaller portions on the left and right of your binocular field where only one eye can see — your monocular field of vision should span about 30 degrees on each side

NOW WHAT?

You just measured your horizontal field of vision, but you also have a vertical field of vision. Try mapping out this out by keeping your eyes focused on an object right in front of you and moving your thumb up and down until you can no longer see it. Is your vertical visual field bigger above or below?

CHAPTER 8

WHY DO

WE FEEL

PAIN?

You're at the doctor's office again. This time it's because you were playing with your sister and you tripped over a tree root in your backyard. You're not sure how that happened! One minute you were dashing towards the back fence that served as "home base," and the next minute you were on the ground next to your favorite climbing tree. Your sister thought you were ok, so she continued running after you and tagged you. That's when you tried to stand up. You felt a jolt of pain travel from your foot all the way up your legs and you collapsed back to the ground. You looked down past your dirt-stained knee to see your very large ankle. It was starting to turn blue and you started to cry because it hurt so bad! Your dad came outside to see what was going on. When he saw you on the ground, he immediately came over and picked you up. You tried to walk on it, but it really, really hurt. You hobbled over to the car and went to the doctor's office for the third time this month. You start to wonder why the pain didn't start the moment you fell to the ground.

FEELING PAIN

Pain is horrible. No one likes to be in pain and we try to avoid it at all costs. Sometimes, however, pain is unavoidable. We have to get our immunizations to protect us from diseases and sicknesses. We fall sometimes when we are learning how to skate or ride a bike. But pain can actually be a good thing. It signals to us that something is wrong and that we better pay attention to it. If it weren't for pain, you might never know if your appendix had burst or that you had an ear infection. These and other ailments are painful, but without immediate treatment, they could be life-threatening.

How much pain someone feels is dependent on a variety of factors. You might think the most obvious factor has to do with what caused the pain in the first place. If you hit your thumb, instead of a nail, with a hammer, that would hurt! But the same amount of force might hurt more if, instead of you hitting your thumb, someone else had dropped a hammer on your hand. *How* you get hurt is just one factor that influences how much pain you feel. *Why* you got hurt also affects how much pain you're in. If someone harmed you on purpose, it might hurt more than if someone did it by accident. If your friend accidentally dropped the hammer and they were really sorry, your pain perception might not be as bad as if they had done it on purpose.

In addition to the physical factors that lead to pain perception, there are psychological factors that influence how much pain you experience. These factors include things like how much sleep you've had lately, the type of mood you're in, and your past experience with pain. Sleep deprivation and poor moods will exaggerate the pain you feel (Chapter 6). This is why stubbing your toe the morning after a late night hurts so much more than if you got a full night's sleep the night before. In addition, if you're used to being in pain, it might hurt less. If you broke your leg once, it might not hurt as much the second time around (but let's not test this theory).

GATE CONTROL THEORY OF PAIN PERCEPTION

It has long been clear that our perception of pain has both physical and psychological causes. This led researchers to hypothesize the **Gate Control Theory of Pain Perception**. In this theory, a pain gate controls whether or not a pain signal is sent to your brain. The pain gate can be closed by non-painful factors that either come from the body or from the brain.

Think back to the last time you accidentally bumped your knee hard on the end of a table. It hurt, right? Now try to remember the first thing you did—after you yelled OUCH really loudly! You probably reached down and rubbed your knee real hard. The pain should have started to go away. This is because when you were rubbing your knee, you were activating your touch sense and that touch signal closed the pain gate. This is an example of a non-painful factor that comes from your body. Now think back to your last vaccine shot. Since you can't rub the area while you're getting the shot, before you get the shot, you can close the pain gate by breathing deep and thinking of being in a happy place. This should make the shot a lot less painful. These are examples of non-painful factors that come from your brain.

THE TIMING OF PAIN

So why don't you feel pain at the exact moment when you get hurt? In an emergency situation, such as the one you might be in when you get hurt, your body activates its **sympathetic nervous system** (more on this in Chapter 19), which gets your body ready to act fast. One thing your body does is to release a hormone known as **adrenaline** (also called **epinephrine**). Adrenaline is released by the adrenal glands, just above your kidneys near your stomach.

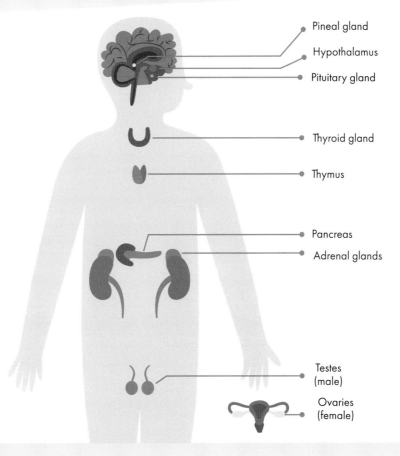

Pineal gland

Hypothalamus

Pituitary gland

Thyroid gland

Thymus

Pancreas

Adrenal glands

Testes
(male)

Ovaries
(female)

Whenever you have a strong emotion like fear or anger, adrenaline is released into your bloodstream. This leads to an increase in your heart rate and an enlarging of your blood vessels so that oxygen and nutrients can be transported to far off areas of your body that may need them to run away or fight off an enemy. Adrenaline is also what allows you to become distracted by other things when you get hurt. In the example from the beginning of this chapter, adrenaline is what led to you not notice the pain in your ankle when you fell to the ground—you were still immersed in your game of tag and so you didn't notice the pain until you tried to stand up.

PICK YOUR BRAIN

Some people are not capable of feeling pain at all. This is called **congenital analgesia** and is extremely rare. While this sounds amazing, this condition is extremely dangerous because being able to feel and know you're in pain is important for survival. These people cannot feel if they've been burned or if they have broken a bone. This is something that people are born with and is due to a genetic mutation which results in the inability of pain neurons in the spinal cord to produce action potentials. If these neurons cannot produce an action potential, then a pain signal cannot be sent to the brain and no pain is felt.

Since everyone feels pain differently, it's really hard to understand just how much pain someone is in. When you go to the doctor, it's important for them to know how much pain you're in so that they can see whether or not the treatment you're receiving is working. One way that doctors can measure pain perception is to use a **visual analogue scale**, which uses faces with various expressions. The patient then chooses the face that most closely matches their pain level.

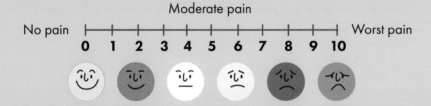

FINAL NOTES FOR YOUR NOGGIN

No one likes to be in pain, but it is a useful signal. There are many different factors that influence the level of pain your feel. How and why you got hurt, your level of sleepiness, your mood, and the intent of the person who hurt you all matter. The fact that pain has both physical and psychological causes led researchers to hypothesize the Gate Control Theory of Pain Perception. This theory states that there is a pain gate that can be closed by signals coming from either the body or brain. Rubbing a mosquito bite can make the pain go away, as can just thinking about something other than the pain or breathing deeply.

PAIN IN YOUR MEMBRANES

Scientists can study pain perception in the lab using a simple task called the Cold Pressor Task. The Cold Pressor Task has been used on children since 1937. In this task, a person places their hand in a cold bucket of water and they slowly start to feel pain from the cold. When the person can no longer tolerate the pain or when one minute has passed, the person pulls their hand out of the water. This task has been used to study the effectiveness of pain medications and interventions.

MATERIALS NEEDED:

- Large bowl
- Ice
- Water
- Stopwatch
- Your hand
- The visual analog pain scale included in this chapter
- Your Brain Journal

STEPS:

1 Ready your Brain Journal by making two columns. Label the first column "Time" and the second column "Pain Level."

2 Record your baseline level of pain by placing a 0 in the Time column and record your pain level using the visual analogue scale given on page 70 in the Pain Level column.

3 Fill the bowl with water and ice.

4 Place your hand in the ice cold water and start the stopwatch. Keep your hand in the water for as long as you can OR until one minute has passed.

5 Rate your pain on the visual analogue scale every 10 seconds.

6 Once you take your hand out of the water, keep on rating your pain level every 10 seconds until your pain level goes back to the baseline pain level.

WHAT HAPPENED?

Your pain level should have been close to 0 before you started. As time went on with your hand in the icy cold water, it should have slowly gotten more and more painful. After you took your hand out, the pain in your hand should slowly get better until you are back down to pre-you-just-put-your-hand-in-icy-water-and-it-hurt levels.

NOW WHAT?

Pain perception is something that varies widely between people. Try asking others to do the Cold Pressor Task and see how their pain perception changes compares to your own—are they quicker or slower to reach the same level of pain? Are there any differences in pain perception between boys and girls? Between young and old people? How does your own pain perception change as you do this task more than once? Or if you do this task when you're in a good or bad mood? Or when you have not had a lot of sleep? Finally, can you try to make the pain feel less by breathing deeply or meditating while your hand is submerged in the icy water?

CHAPTER 9

WHY DOES LEARNING HOW TO RIDE A BIKE SEEM SO HARD AT FIRST?

WHEEEEEEEEEEEEEE!

You just got a new bike and you can't wait to take it out for a spin around the neighborhood. You put on your helmet and you ride over to your friend's house to show him your new wheels. He's pretty excited about it since it means you two can now ride bikes to the local grocery store that sells candy and soda. He runs inside the house to tell his dad that you two are going to Larry's Market, but when he comes back, he looks upset. His dad told him he can only go if he takes Finn, his younger brother...who doesn't know how to ride a bike. So, that means if you're going to be able to get your hands on candy and soda, you're going to have to walk. Not wanting to spoil your good mood, you suggest that you and your friend teach Finn how to ride a bike. It can't be that hard. You've been riding a bike for as long as you can remember. All you have to do is sit on the seat, pedal, and steer. In fact, that's what you tell Finn, but when he tries to do that, he just falls. You encourage Finn to get back on the bike and you hold onto the back while he starts pedaling. As soon as you let go, though, Finn falls again and then refuses to get back on. You wonder why riding a bike seems so hard for Finn.

YOUR MEMORY

There are certain things you just know how to do, even if you can't explain it. Riding a bike, tying your shoes, climbing up stairs, and even walking are some examples of things you do without thinking much about them once you've learned how to do them. These learned motor skills are part of your **procedural memory**. Procedural memories are those activities you can do without much thought. Not only do you do them without much thought, it's really hard to explain *how* you do these things after you learn to do them. That is why it's so hard to teach someone else how to ride a bike. You can tell them things like sit, pedal, and steer, but riding a bike involves so much more. You may not realize it, but your body is making all sorts of small motor adjustments to keep yourself on the bike. These small adjustments are not something you're conscious of (see Chapter 12), but they are an important part of the learning process.

A procedural memory is a type of **long-term memory**. Long-term memories are those that you store for long periods of time, like learning how to ride a bike, remembering what you did on your last birthday, and knowing the names of all 50 state capitols that you learned in fourth grade. Procedural memories, like learning how to ride a bike, are a kind of **implicit long-term memory.** Implicit memories are those that you cannot explain where you learned them or how to do them. **Explicit long-term memories** are those you can *explicitly* describe and talk about. Explicit memories about particular instances in your past, like your last birthday, are called **episodic memories**. Memories about facts that you've learned, like the 50 state capitols, are called **semantic memories** (even if you can't quite remember them without someone giving you a clue).

If long-term memories are those that you remember for a long period of time, **short-term memories** are those that you remember for a short period of time. Remembering a list of items to pick up at the grocery store is an example of a short-term memory. If you want to remember something for a longer period of time, you need to spend some time studying the material (Chapter 4), or maybe even sleep on it. In fact, scientists believe that one reason why we sleep is so that we can convert short-term memories into long-term memories (Chapter 6).

PROCEDURAL MEMORIES

Procedural memories form very early in life. When you are born, you need to learn all sorts of motor skills, including how to walk, sit up, and talk. You learn how to do all of these things through interacting with your environment.

Experience is really key here because procedural memories are formed through trial and error. Take learning how to ride a bike. It takes many attempts, and probably many scraped knees, to learn how to ride a bike properly. Procedural memories are formed based on your personal experience. This is why you can't just tell someone how to ride a bike—they have to experience riding the bike firsthand, bumps and all.

When you complete the same actions over and over again, you are strengthening the connections between the groups of neurons that are involved in the actions. This makes the behavior change from something you have to think about constantly to one that becomes automatic. When you first learned to ride a bike, you had to be aware of where you were steering, how fast or slow you were pedaling, and how you needed to shift positions to remain steady and balanced on the seat. Once you learned how to ride the bike, you could just sit back and enjoy the ride! The behavior became automatic.

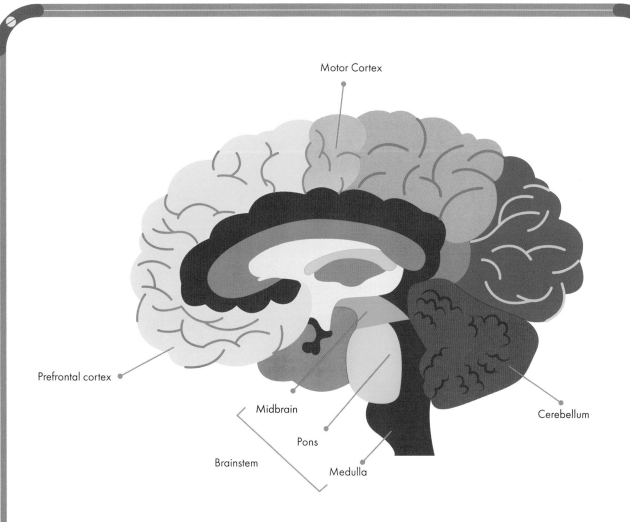

Motor Cortex

Prefrontal cortex

Midbrain

Pons

Brainstem

Medulla

Cerebellum

There are many brain areas that are involved when learning how to ride a bike. The **cerebellum** plays a huge role in many of the different motor skills that you've mastered, including walking. The cerebellum is the brain area that coordinates the flow and timing of all your movements. People who suffer damage to the cerebellum are not able to move about smoothly, if they are able to move at all, although physical therapy may help (Chapter 23). Another area involved in procedural memories is your **motor cortex,** which is located in your **parietal cortex**. The motor cortex is the brain area that sends signals to your muscles to move. Finally, areas such as the **prefrontal cortex** and structures in the **limbic system** are both involved and are important for learning and memory.

HM is a man who had both of his temporal lobes surgically removed because in 1953 he suffered from severe epilepsy. While the surgery partially cured his epilepsy, it left him with a devastating side effect. HM could no longer form any new explicit memories. At first, scientists believed that he could not form any new memories, but it turns out that he was still able to form new implicit memories. For example, he could learn how to play games (an implicit memory), but he could not remember ever having played them before (an explicit memory). This led scientists to the idea that explicit and implicit memories were created and stored in different areas of the brain.

FINAL NOTES FOR YOUR NOGGIN

Learning how to ride a bike is an example of a type of implicit long-term memory called procedural memory. It's difficult to clearly explain how to do procedural memories. There are also explicit memories, which are those you can talk about in clear terms, like your last birthday party or the 50 state capitols. Many brain areas are involved in procedural learning, including the cerebellum, motor cortex, prefrontal cortex, and the limbic system. Procedural memories are learned through personal experience and trial and error. The more times you practice a skill and get feedback about how well you are performing that skill, the stronger the connections between brain areas become. This leads to behavior becoming automatic and not something you're actually conscious of doing.

SIGN WRITE HERE

In this Lobe Lab, you will try your hand at learning a new procedural memory. Learning how to write is an example of a procedural memory. You probably don't even remember when you first learned to write, and while you can tell others how to spell your name, you can't describe how you personally sign your name. A person's signature is unique to them and it's something that is not easy to teach someone else how to do, unless they are practicing the signature over and over again.

MATERIALS NEEDED:
- A friend or family member
- Your Brain Journal

1 Make up a name, like Everett Engineer, and write his name in your Brain Journal 10 times. You want to make sure you can write the name so that it looks more or less the same every time you write it.

2 On a different page in your Brain Journal, have your friend close their eyes and write out Everett Engineer's name as you describe the signature. Make sure you describe every part of every letter and the transitions between the letters. Have your friend do this 10 times. Compare and contrast how the signatures look the same or different in your Brain Journal.

3 Make up a second name, like Zoe Zoologist, and write her name in your Brain Journal 10 times. Again, you want to make sure you can write the name so that it looks more or less the same every time you write it.

4 Have your friend write out Zoe Zoologist's name as they look at the name as you have written it. Have your friend do this 10 times. Compare and contrast how the signatures look the same or different in your Brain Journal.

WHAT HAPPENED?

At the end of the 10th time writing the name, your friend's copy of Zoe's signature should look closer to the original than their copy of Everett's signature. This is because your friend was actively engaged in figuring out which parts of the signature were off from the original and was able to make small adjustments according to what they were seeing. If you ask your friend why they did better with Zoe's signature, they might not even be able to explain why they did much better. This is because procedural memories are not easily described—it's just as hard to teach someone how to write something as it is to teach them how to ride a bike.

NOW WHAT?

The longer a person knows how to do something, or the more they engage in the activity, the stronger the neural connections will be for the activity. The strength of the connections is related to how long a person can learn a new, related skill. Ask an adult or a younger friend to do the same activity as above. See if you can find a relationship between how long a person has been writing (by using their age) and how many times it takes for them to learn how to copy a signature almost perfectly.

CHAPTER 10

WHY DO ICE CREAM HEADACHES HAPPEN?

YOU SCREAM, I SCREAM, WE ALL SCREAM FOR ICE CREAM

It's the middle of summer, and you're excited about the county fair that is in town for the week. You and your friend ride your bike to the fairgrounds and purchase an all-day ticket so you can ride all the rides! You start out riding the scrambler and then walk your way through the fun house. Then you slide down the big slide and ride the Ferris wheel. As you are sitting on top, you look around at all the lights from the various fair games and food vendors and then you spot it! You knew the Incredible Ice Cream Man was going to have a food truck at the fair and now you know he's right by the balloon popping game. You and your friend hop off the ride and head straight for the ice cream. You order an extra-large chocolate chip peanut butter fudge ice cream in a waffle cone. You pay for your little bit of heaven and sit on the plastic chairs next to the truck. The first few bites are fantastic! Then all of a sudden—OW! The pain in your head makes you feel like your brain is about to explode. Such is the curse of an ice cream headaches. It is technically known as a *sphenopalatine ganglioneuralgia*, and if you say that five times fast, you'll get more than an ice cream headache!

I. Olfactory	II. Optic
III. Oculomotor	V. Trigeminal
IV. Trochlear	VII. Facial
VI. Abducens	IX. Glossopharyngeal
VIII. Vestibulocochlear	X. Vagus
XII. Hypoglossal	
XI. Accessory	

If you want to understand why you feel something, you have to understand how your nerves work. You have 12 pairs of **cranial nerves**, which connect your brain with the rest of your body. You can see all 12 pairs if you flip the brain upside down and look on its **ventral** surface. The cranial nerves (CN) are numbered 1–12 (in Roman numerals I-XII) in order, starting from the **rostral** (front) end of the brain to the **caudal** (back) end and you have one on each half of your brain. Each pair of nerves serves a special purpose. Some of the nerves allow you to experience the world with your five senses. For example, your Olfactory Nerve (CN I) allows you to smell pretty flowers in the garden...and disgusting, rotten food in the garbage. Other nerves carry information from your brain to your muscles in order for you to move. For example, the Oculomotor Nerve (CN III) allows you to look around on a hike to see what animals are hiding in the trees. Finally, other nerves, like the Vagus Nerve (CN X), are connected to your internal organs like your heart so that you can feel it racing when you have to give that presentation in front of your classmates.

FLAVOR PERCEPTION AND YOUR CRANIAL NERVES

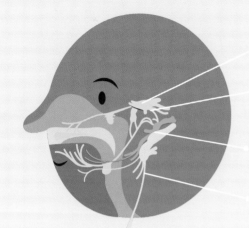

Trigeminal Nerve
(CN V)

Facial Nerve
(CN VIII)

Glossopharyngeal
Nerve (CN IX)

Vagus Nerve (X)

So, what do cranial nerves have to do with eating ice cream? Quite a bit! To understand how, we need to talk about what factors influence your flavor perception. When people say that something *tastes funny*, they are actually referring to the *flavor* of the food. Flavor is a complex thing made up of all the other senses. Taste is just one aspect of flavor. Think about how your other senses influence your opinion about a food's flavor. Would you eat:

| Vision: a blue banana? | Hearing: a crunchy strawberry? | Smell: a stinky meatball? | Texture: a slimy chicken nugget? | Temperature: a cold bowl of vegetable soup? |

Most people would not enjoy eating those foods, but they might eat a blue plum, a crunchy taco, stinky blue cheese, a slimy olive, and a cold bowl of ice cream. It's not that there's anything wrong with blue, crunchy, stinky, slimy, or cold food outright. It's that we have certain expectations about certain foods. When those expectations are not met, it influences your flavor perception, and you think, "GROSS!"

You have many cranial nerves that are in charge of processing information about the flavor of the food you eat. The cranial nerves get their information from the taste buds that are in structures called **papillae** on your tongue. Three of the nerves process information about the taste of the food. The facial nerve (CN VII) gets information from the front two-thirds of your tongue from the fungiform and foliate papillae and the glossopharyngeal nerve (CN IX) gets information from the last one-third of your tongue from the foliate and circumvallate papillae. The vagus nerve (CN X) also gets taste information from the back of your mouth. Finally, your trigeminal nerve (CN V) is involved in flavor perception, but not taste perception. The job of the trigeminal nerve is to tell your brain about the texture, temperature, and feel of the food you put in your mouth.

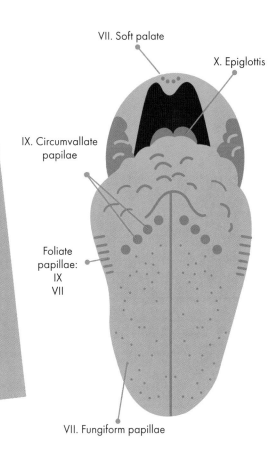

VII. Soft palate

X. Epiglottis

IX. Circumvallate papilae

Foliate papillae:
IX
VII

VII. Fungiform papillae

EATING ICE CREAM

Now that we understand more about how your cranial nerves all work together so that you can taste the flavor of food, what does feeling the pain of an ice cream headache have to do with any of this? When you eat something really, *really* cold like ice cream or a slushie, the cold first constricts the blood vessels in your mouth. Then, as you swallow the cold, yummy substance, the blood vessels dilate. This causes a rapid change in blood flow, which can affect the surrounding nerves and skin around the constricting and dilating blood vessels. These changes are sensed by the trigeminal nerve and cause them to release a chemical called **Substance P**, which sends a message to the brain that is interpreted as PAIN!

Next time this happens, note *where* in your mouth you feel the pain. It is usually in the back on the roof of your mouth. If you want to avoid getting an ice cream headache, try taking small bites and don't let the ice cream touch the roof of your mouth. If you're drinking an ice-cold drink from a straw, try to not place the straw so that its content empties in the back of your throat. Or you could just avoid eating ice cream altogether. Yeah, right!

PAIN AND EATING

Ice cream headaches are not much fun and cause a lot of pain. But part of the enjoyment of eating spicy foods, like chili peppers, includes the pain that is associated with eating these foods. Chili peppers contain a chemical called **capsaicin**, which activates the nerve cells in your trigeminal nerve (V), leading to the release of the neurotransmitter **Substance P** and the perception of pain. However, you chili pepper eaters might notice that after you've been eating chili peppers for a while, they start to not taste as hot. This is called **adaptation**, and it occurs because your nerve cells have temporarily run out of Substance P. As you're eating the chili pepper, your trigeminal neurons are releasing Substance P at a constant rate. Since it takes time for Substance P to replenish, eventually, the nerve cell runs out of it. If there's no Substance P, there's no hot flavor. This means that you're going to have to wait a while to taste the delicious hotness once again!

PICK YOUR BRAIN

Have you ever gone to pick up your drink of water, but accidentally picked up the glass of milk sitting beside it? What did the milk taste like when you were expecting water? Probably pretty gross since it didn't match your expectation of drinking water. Our expectations very much influence how we think something is going to taste. So, next time someone encourages you to try a new food, expect that you will really like it, and you just might!

FINAL NOTES FOR YOUR NOGGIN

Your perception of the flavor of food is made up of more than just taste. How something looks, sounds, feels, and smells influences your perception of how good (or bad) something "tastes." All of this information is sent to the brain via your 12 cranial nerves. When you eat something cold, it rapidly constricts and dilates the blood vessels in your mouth. The brain interprets this as pain. Pain is a part of your eating experience whether it's intentional (chili peppers) or not (ice cream).

EATING FOR FUN

Many people do not realize that the flavor of something is not the same as its taste. Flavor is a complex sensory experience that involves all five of your senses, as well as your expectations about what it's going to taste like. In this lab, you'll get to see how your sense of smell influences your flavor perception.

MATERIALS NEEDED:
- Different flavors of jellybeans—at least 3 different flavors/colors
- Blindfold (optional)
- Your Brain Journal

STEPS:

1 Select at least three different colored jellybeans from the bag.

2 In your Brain Journal, create a simple chart with four columns labelled: "Jellybean color", "What I think its flavor will be", "What it tastes like with my nose pinched", and "What its actual flavor is".

3 Record the color of each jellybean in your chart. Hypothesize what you think the flavor of each jellybean will be.

4 Close your eyes or cover them with a blindfold. No peeking!

5 Pinch your nose so that you cannot smell anything.

6 Without looking, place one jellybean in your mouth and chew it. Try to guess what flavor it is. What does it taste like to you?

7 Before you swallow the jellybean, unpinch your nose. Can you tell what flavor it is now?

8 Open your eyes or take off your blindfold or bandana. Figure out which jellybean is missing—this is the one you just ate! Record what the flavors were with your nose pinched and unpinched in the third and fourth columns of your table.

WHAT HAPPENED?

Did you notice that you could not tell the flavor of the jellybean until your unpinched your nose? The taste of jellybeans and the flavor of jellybeans are two very different things. Our experience of flavor requires not only its taste, but its smell. This is because you need your sense of smell to perceive the flavor of the jellybean. When you unplug your nose, the smell from the jellybeans travels up your throat, to your nose, through the back door. This is called **retronasal olfaction**.

NOW WHAT?

You can try this with a variety of foods, even those you don't like! You'll have to be honest with yourself though and really give this a fair chance. Pinch your nose and try a bite of your least favorite food. Can you taste the food while your nose is plugged? You can also try this trick with nasty tasting medicine—forget the spoonful of sugar, just plug your nose and you won't taste it...until you unplug your nose, that is! Good luck!

CHAPTER 11

WHY

CAN'T I EVER GET THAT ITCH

ON MY BACK?

YOU SCRATCH MY BACK AND I'LL SCRATCH YOURS

You're sitting down by the campfire after a long day hiking and, all of a sudden, you feel it. An itch right in the small of your back. You reach back and try to scratch the area, but you just can't seem to reach it. The itch is really bothering you now, so you get up and ask your brother to help you out and get that itch. You start directing him to the offending area: up, up, to the left, my other left, a little more, too far, down a bit, up just a tad, ahhh, that's it. I think. Why is it so hard to tell someone where that itch is when it's on your back? You feel it, so why can't you locate it? On the other hand, pun intended, if you have an itch on your hand, it is really easy to find and scratch. What is the difference?

YOUR TOUCH SENSATIONS

What you call your sense of touch is actually made up of all sorts of sensations, including pressure, pain, temperature, vibrations, tickles, and itches. Every time you *feel* something, it is registered by **free nerve endings** in your skin and then sent to the brain. Different sensations are registered in your skin by different kinds of free nerve endings, including **mechanoreceptors** for pressure and touch information, **thermoreceptors** for temperature information, and **nociceptors** for pain information (see Chapter 8). Each type of neuron only responds to a specific type of sensation. Imagine you're camping for the weekend. Think about the sensations you feel. Your mechanoreceptors will let you know how soft your sleeping bag is, how hard the forest floor is, and when a mosquito bites you. Your thermoreceptors will let you know how warm the campfire feels and if it's too hot, your nociceptors will let you know that you're in pain and it's time to move back from the fire with your marshmallow on a stick.

SOMATOSENSORY CORTEX

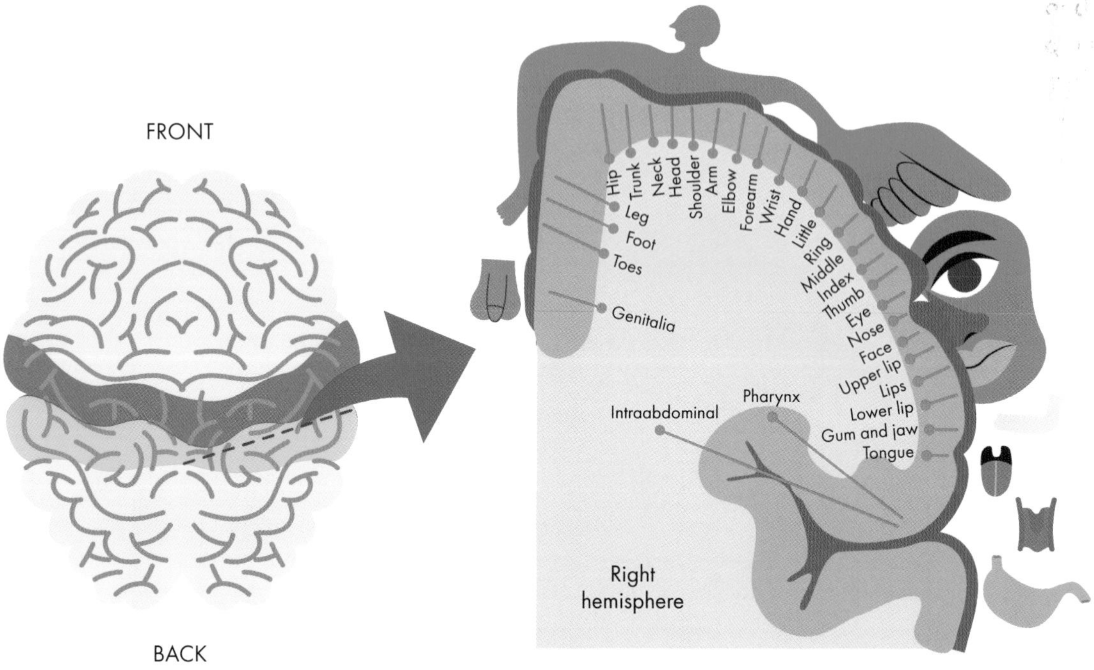

FRONT

BACK

Hip
Trunk
Neck
Head
Shoulder
Arm
Elbow
Forearm
Wrist
Hand
Little
Ring
Middle
Index
Thumb
Eye
Nose
Face
Upper lip
Lips
Lower lip
Gum and jaw
Tongue

Leg
Foot
Toes

Genitalia

Intraabdominal

Pharynx

Right
hemisphere

Most of the touch information from your free nerve endings in your body travels first to the spinal cord and then up to your brain, where that information is interpreted. Did you feel a mosquito on your arm or was it actually a branch that swayed in the wind and poked you? The **somatosensory cortex** is the part of the brain that processes all the sensations that you feel and is located in your parietal cortex just behind the **central sulcus**, the huge valley that separates the frontal and parietal lobes (see Chapter 2). It contains a somewhat orderly map of your body from your toe (located on the top of the brain) to your mouth (located at the bottom of your brain). You actually have two maps, one on each side of the brain for each side of the body. The maps in your brain are **contralateral** to your body parts, meaning your right-side body sensations are processed in your left somatosensory cortex map and vice versa. When you need to figure out if the mechanoreceptor in your right arm was touched by a mosquito or rogue branch, your left somatosensory cortex gets to make the call.

The body map in each hemisphere is called the "homunculus," which is Latin for "little man." Although the map has all body parts represented in your brain cortex, not all body parts have equal representation. For example, the part that represents your hand and face make up about a third of the map, even though these areas make up a much smaller portion of your body. This is called **cortical magnification** because it is like your brain uses a magnifying glass for certain parts of your body and makes them seem bigger in your brain. Bigger brain representations mean that there are more neurons in your brain devoted to processing information from those areas of your body relative to others. Having more neurons process information leads to you being more sensitive to touch information from those parts of the body. Think back to that magnifying glass. If you look at a rock through a magnifying glass, you can see a lot more texture from the rock than if you were just viewing it with the naked eye. In the same way, you feel more from a body part that is magnified in the brain.

Back to that itch on your back. Even though your back is a large part of your body, its representation in your brain is not that large, so you're not as sensitive on your back and it's hard to tell where the itch is coming from. Your hand has a large representation, so you can tell precisely where that mosquito has bit you so you can scratch the itch away!

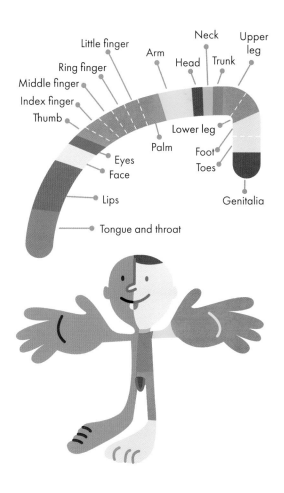

THE HEALING POWER OF TOUCH

Don't feel too bad for your back! Even though it is harder to find an itch on your back, it can still feel the warmth of a hug or a reassuring pat on the back. Turns out, touch is a powerful healer. Think about the last time you were upset. Did someone offer you a shoulder to cry on, or touch your arm or shoulder gently to let you know that things were going to be okay? You probably felt better afterwards because a simple touch can lessen one's pain and anxiety. Touch can also increase your immune system's capability to fight off infection, lower your blood pressure and the production of stress hormones, and improve circulation. Touch can also stimulate human growth! It used to be that premature babies were kept in isolation in covered cribs that protected them from infections from the outside world. Unfortunately, this also meant that the infants could not be touched. Doctors and scientists discovered that babies would fare much better if they were held and if their backs were stroked gently. Babies who are touched grow up to be less aggressive and more confident. So, go ahead and give your good friend a hug if they're upset —with permission of course!

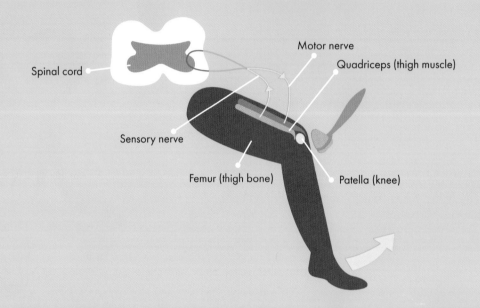

PICK YOUR BRAIN

Imagine you are at the doctor's office for a checkup and she hits your knee with her hammer. Your foot seems to kick up, all on its own, doesn't it? This is called the **knee-jerk reflex**. It happens because when the hammer hits your knee, it stretches your knee muscles. That information gets sent to the spinal cord and, without consulting with the brain, the spinal cord sends a signal back to the knee muscle to contract. So, now you have an excuse if your doctor stands in front of you when you kick up—you didn't decide to kick them, your spinal cord did!

Spinal cord

Motor nerve

Quadriceps (thigh muscle)

Sensory nerve

Femur (thigh bone)

Patella (knee)

FINAL NOTES FOR YOUR NOGGIN

Touch sensations are actually made up of different kinds of touch like pressure, pain, temperature, vibrations, tickles, and itches. These different touch sensations are registered by different types of free nerve endings. Touch information from the body is sent to the somatosensory cortex where there are maps of body space. These maps are not true representations of the body because certain body areas, like your hands, have greater representation in the cortex compared to the size of the body part. This is called cortical magnification. Parts of the body that have smaller representations in the brain are not as sensitive, like your back, which is why it's hard to get that itch on your back.

CAN YOU FEEL IT?

In this Lobe Lab, we're going to measure just how sensitive your arm is relative to your back. You can work with a friend, sibling, or parent. You should take turns gently poking each other with one or two sticks. You will have to decide whether there are two sticks or one stick touching you. The trick is that you'll be changing the distance between the two sticks. The more sensitive you are, the smaller the distance between the two sticks will be for you to tell that there are two sticks, and not one stick. This is called your **two-point threshold.**

MATERIALS NEEDED:
- 2 wooden sticks with pointed ends, or math compass
- A friend
- Blindfold (optional)
- A ruler that can measure in centimeters
- Your Brain Journal

STEPS:

1 Create a table in your Brain Journal with three columns. Label the first column, "Part of Body" and then write down the body part you test in that column. Label the second column, "Distance" where you will record the distance between the two sticks. Finally, label the third column "Response" where you will record whether your friend feels one or two points.

2 The person being touched should cover their eyes with a blindfold, or simply close their eyes.

3 Start with either the arm or back. Put the two sticks pretty far apart and ask your friend whether they feel one or two sticks. Note how far apart the two sticks are located and record that and their answer in the table.

36 to 75 mm

4 Keep repeating the process, but either move the sticks together a little more or only put one stick on their skin to make sure they can't guess whether you are putting one or two sticks on their skin. Note the distance each time you put two sticks on their skin and record it in the table.

3 to 8 mm

5 Stop the process when your friend says they feel one point. This is the two-point threshold for that body part.

6 Repeat this process for the other body part and measure its two-point threshold.

1.1 mm

WHAT HAPPENED?

The two-point threshold should be bigger on the back compared to the arm. This is because your arm has a bigger representation in your brain than your back. Body parts that have a bigger representation are more sensitive to touch. One of the most sensitive parts of your body is your lip. You can try to measure the two-point threshold there, but you probably can't get the two points close enough for your friend to say one point because it's very sensitive. Also, your friend might not like that!

NOW WHAT?

You just measured the two-point threshold for your arm and back. Try repeating this experiment with different body parts. What part of the body do you think is most sensitive? What part of the body do you think is least sensitive? In other words, where on the body can two points be detected with the smallest tip separation?

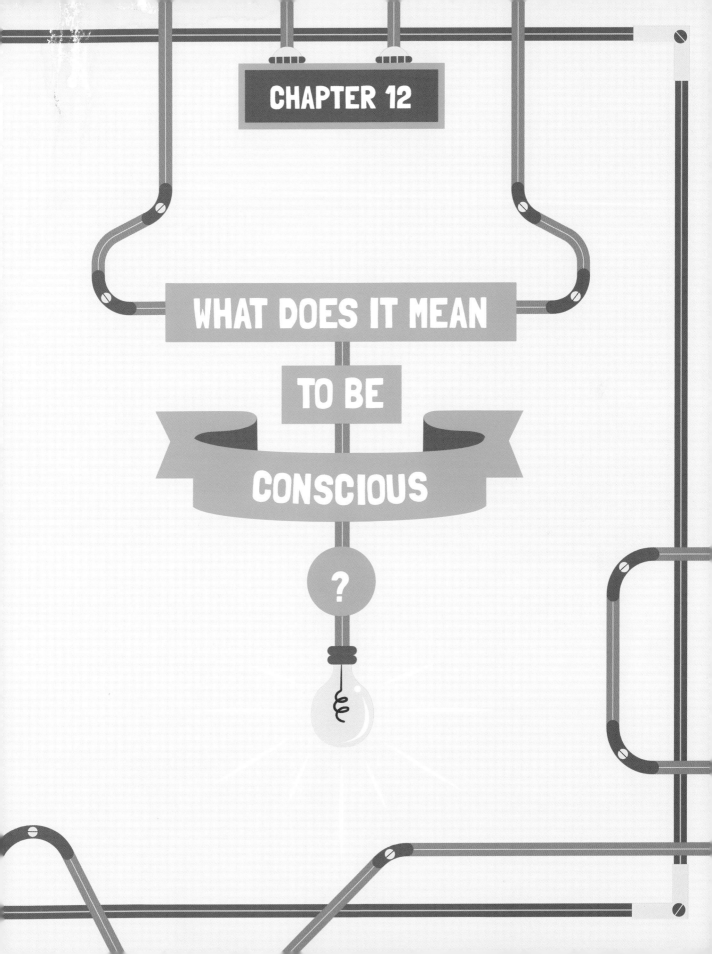

CHAPTER 12

WHAT DOES IT MEAN

TO BE

CONSCIOUS

?

PAY ATTENTION!

There you are sitting in class, daydreaming away. You're thinking of the big game this weekend where you are finally going to be able to play second base. You watch the pitcher throw the ball and hear the crack of the batter's bat and start to anticipate which way you will have to lunge in order to catch the ball. All of a sudden, you hear your name. The crowd is going wild you think! They are really rooting for you! But you hear your name again and there you are, back in class. "Pay attention!" the teacher yells at you. "Okay, okay," you think, but you're a bit annoyed that you've been brought back to reality. Reality, you think, now that's a funny concept. When you were daydreaming, playing baseball felt like reality to you, more so than sitting in this class. You thought you even smelled the freshly cut grass and the newly laid red clay beneath your cleats. You start to wonder what reality is and what it means to be conscious of it.

DEFINING CONSCIOUSNESS

In order to study anything, it needs to be well defined. After all, you can't study how an *eonatric* works until you know what it is (by the way, *eonatric* is a made-up word!). This is the problem with studying consciousness—it is hard to define. The topic has inspired debates and arguments among philosophers and scientists for centuries. Most modern theories define consciousness as what you are aware of, including your thoughts, memories, and interactions with the environment.

Given that definition, the question becomes how do you study a person's awareness? Early psychologists used the technique of **introspection**, where people described their experiences as a way of measuring one's conscious life. Our first-hand experience of the world is what scientists call **phenomenology**. The problem with introspection is that there is no way to verify what it is you are experiencing. Consider your experience with the color blue. How do I know what you experience as blue is the same experience as I have with the color blue? Sure, we may be able to sort colors into blue and non-blue categories, but how do I know that what you see as blue is the same thing I see as blue?

How you see the color blue is a phenomenological question. On the other hand, how you interact with the color blue is an **empirical** (observable, testable) question. Modern psychologists and neuroscientists use empirical methods, including behavioral responses (like how you might sort colors) and brain activity (like which areas are active when you are deciding whether something is blue or non-blue), to study our conscious experience.

A philosopher named David Chalmer distinguished between the easy and hard problems of consciousness. The **hard problems of consciousness** have to do with answering the **mind-body** questions. That is, how does neural activity give rise to our experience of the world? For example, when we say that the "visual cortex is active when we are looking at something," how do the action potentials in the brain (i.e., the *body*) turn into our visual perception (i.e., our *mind*)? You can see how this is a hard problem to solve. The **easy problems of consciousness** have to do with answering questions of how we pay attention to, discriminate between, and integrate information from the many different sources of information we get from our environment. The easy problems of consciousness look at both brain activity and behavioral responses to answer these questions. Once we understand how the brain does this, the easy problems are "solved."

STUDYING CONSCIOUSNESS

Some scientists study the easy problems of consciousness by examining how people move through different states of consciousness. For example, they may study brain activity or behavior while a person goes from being fully awake to asleep (Chapter 6). Other scientists study how much of our everyday lives happens in the *absence* of our awareness.

You may not be aware of it, but there is plenty that happens during the day that you are just not aware of because you are not paying **attention** to it or because you have become **adapted** to it. Attention is when we selectively concentrate on something at the expense of other things. For example, in a crowded restaurant, there are many people talking around you, but you can still have a conversation with your friends by selectively listening to them and ignoring all other conversations. When you become adapted to something, you start to pay less attention to it. For example, when you first sit down, you feel the chair hit your bottom. But after you've been sitting for a while, you barely notice the feel of the seat on your butt.

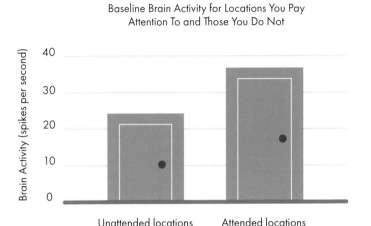

Baseline Brain Activity for Locations You Pay Attention To and Those You Do Not

Brain Activity (spikes per second)

40
30
20
10
0

Unattended locations   Attended locations

Neuroscientists study attention and adaptation by examining how brain areas become more or less active in a particular situation. Activity level is important because most areas have to reach some threshold level of activity in order for a decision to be made or an action to be performed. Every brain area has a particular baseline level of activity; the brain is never silent. If you can increase that baseline level of activity by paying attention to something, the higher level of activation can lead to faster decision and response times. If you can decrease that baseline level of activity by becoming adapted to something, the lower level of activation can lead you to either respond much slower, or not at all.

For example, let's say you are looking for your friend in a crowded room and have to guess whether they will walk through the door on the left or right of the room. If you are told that your friend will probably walk through the right door, the neurons that are processing the right area of space will have a greater level of baseline (resting) activity (as shown on the right side of the above graph) than those processing the left side of space (as shown on the left side of the graph). It's as though your neurons are getting ready to respond quicker to your friend! And it turns out that not only is your neural activity ready to respond quicker, you react faster if you are paying attention to something. In this case, you may be able to walk faster toward your friend if they come through the right door than if they come through the left door.

On the other hand, when you become adapted to a stimulus, the neural activity becomes less. Think about the last time you walked into a house (or bakery) where someone had just made homemade chocolate chip cookies. The smell is glorious, and your mouth starts to water because you can't help but think about taking a bite out of that warm gooey cookie. But after a few minutes, you realize that it doesn't smell as strongly, if at all. You start to wonder why you ever thought it smelled so good. It's not the cookies that are smelling less, it's that you've become adapted to the smell. The smell isn't changing—it's not becoming stronger, weaker, or even different. It's staying the same. And because it's staying the same, your nervous system has decided to pay less attention to it. This is because the neural processing of signals is energy intensive, and if nothing is changing, then there is no need to expend energy on analyzing the incoming signals.

Did you know that you have a blind spot in each of your eyes—a spot where you don't see the world? You have a blind spot because there's a spot on the back of your eye where there are no light-sensitive neurons. Usually, you are not aware of this blind spot because your brain fills in the missing information, but you can find the blind spot yourself by doing the following: close your left eye and stare at the circle. Move the book toward and away from you until you no longer see the X. When the X falls in your blind spot, your brain fills in the space with the surrounding visual world, in this case, a blue background.

William James, commonly referred to as the father of American psychology, was the first to coin the term "stream of consciousness." It was meant to be a metaphor for how we experience life as a never-ending stream of thoughts.

FINAL NOTES FOR YOUR NOGGIN

Consciousness is difficult to understand because it is hard to define. David Chalmer divided the question of consciousness into two sets of problems: the easy and the hard. The hard problems of consciousness have to do with answering questions about how neural activity leads to our perceptual experiences. The easy problems of consciousness have to do with solving questions about how the brain pays attention to the world in order to discriminate and integrate information from many different sources. Neuroscientists can study how people's behavior and their brain activity is changed by how much attention they are paying (or not paying) to something in the world. Both brain activity and behavior are enhanced when you pay attention to something, and both are diminished when you become adapted to the same something.

DUELING EYES

Our brain is constantly making decision s about what to pay attention to in our visual world. Our brains do not like indecision and inconsistencies. That is why it "fills in" missing information in our blind spot. Your brain makes a best guess about what was supposed to be in the blind spot since there is no neural activity indicating what was actually there. The problem becomes more complicated when there is a competing stimulus for the same part of the visual world and the brain has to decide which neural activity to pay attention to. This is called **binocular rivalry** and it's when the two eyes are sending the brain two different visual signals. The brain has to figure out which eye to pay attention to and which eye to ignore. In this Lobe Lab, we're going to intentionally send two different visual signals to the brain and witness as the brain tries to figure things out!

MATERIALS NEEDED:

- A cardboard tube or a rolled-up piece of paper
- Your hand
- Your Brain Journal

STEPS:

1 Draw something fun (like a neuron with a silly face on its soma) in your Brain Journal.

2 Close your right eye. Prop up your Brain Journal and look through the cardboard tube with your left eye so that you can see your silly drawing through the hole in the tube.

3 Place your open right hand next to the cardboard tube with your palm facing towards you.

4 Open your right eye and try to relax both your eyes and look.

5 Draw or write what you observe in your Brain Journal. Why do you think you see this?

WHAT HAPPENED?

After a while, you should see your silly drawing start to creep over your hand and look as though it's part of your hand. This is because your brain is trying to combine the signals from both eyes as best it can. In this case, the brain is getting the input of your silly drawing from the left eye and the input of the middle of your palm from your right eye. To solve this problem, the brain ignores the input from your right eye in favor of the input from your left eye. This isn't a permanent solution though. What you see will oscillate between the inputs from the left and right eyes, with the left eye "winning" sometimes and the right eye "winning" other times.

NOW WHAT?

Look through two cardboard tubes, one for each eye. You should see just a circular slice of your visual world surrounded by the tube that is different for each eye. It will be hard for your brain to ignore one eye's input and just see one thing, so you'll go on seeing two different things from both eyes. Now just use one tube again. Close your right eye and look through the tube with your left eye. While looking through the tube with your left eye, open your right eye to look normally (without a tube). The sides of the tube will disappear, and you should see the whole visual world in front of you. Your brain is still getting two different signals: your left eye is sending the signal of your silly drawing surrounded by the tube, but when you open your right eye, the tube "disappears." This is because the whole visual world is much more visually appealing than a blank space so it will be the more dominant input and will "win" the binocular rivalry contest a lot more frequently.

CHAPTER 13

WHY CAN'T WE HEAR

DOG

WHISTLES?

HERE, DOGGIE!

Well, it happened again. You opened the front door to get the mail and your dog, Sadie, ran right out the front door, down the front walkway, and down the street. You start running after her yelling, "Sadie! Come here, Sadie!" but she just ignores you and runs away. You trudge back into the house to get her dog whistle. You trained her using classical conditioning (Chapter 4) to associate the sound of the whistle with her favorite bacon flavored treat. You run outside to the end of the sidewalk and blow into the whistle real hard. You can't hear it, but you know Sadie can. You blow it three more times, but still Sadie refuses to answer. You start to wonder if the whistle is broken. You try one last time, and you see Sadie come running from outside your neighbor's backyard straight to you. You walk fast back into the house, blowing the silent-to-you whistle the whole way. Once Sadie gets inside, you shut the door and head to the kitchen to give her a treat. You marvel at how awesome it is to have such a whistle, and you start to wonder why Sadie can hear it, but you can't.

SOUNDS

Have you ever heard the question, if a tree falls in the woods and no one is around to hear it, does it still make a sound? Neuroscience has the answer to that question—it does indeed depend on whether or not someone or something was around to hear it! Sound is nothing but air waves that are *captured by your ears* and processed in your nervous system. Air waves would be created by the fall of the tree, but they would not be processed into sound without your ears! Sound waves have certain physical properties that lead to different perceptual qualities. The physical qualities of sound waves, their **amplitude** (how big) and **frequency** (how fast), lead to the perceptual qualities of loudness and pitch perception. Imagine splashing around in a pool with your sister. If you shove a whole bunch of water her way, you can make a tall wave, or one with a large amplitude. Loud sound waves are like tall waves, and soft sound waves are like short waves. Now imagine you keep pushing the wave toward her at a steady slow rate, or at a low frequency. Sound waves that ebb and flow at a slow rate are low in pitch, like what the distant sound of thunder sounds like. Fast moving sound waves are high in pitch, like the squeal your sister makes when you splash her. Loudness is measured in decibels (dB) and pitch is measured in Hertz (Hz).

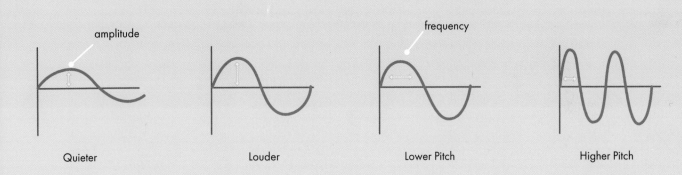

Not all sound waves can be heard by humans. In fact, there are sound waves all around that you probably can't hear. Humans can hear pitches that are between 20 and 20,000 Hz. A dog can hear a dog whistle because dogs can hear much higher pitches than humans. A dog can hear pitches between 64 and 44,000 Hz. The pitch of a dog whistle is typically above 23,000 Hz, so the dog can hear it, but humans cannot. On the other end of the spectrum, an elephant can hear sounds that are lower in pitch than you can. An elephant can hear pitches between 16 and 12,000 Hz. But since sound waves are physical things, you can sometimes feel low-pitched sounds, even if you cannot hear them. A good example of this might be feeling an earthquake that happened far away from you, even if you can't hear it.

THE AUDITORY SYSTEM

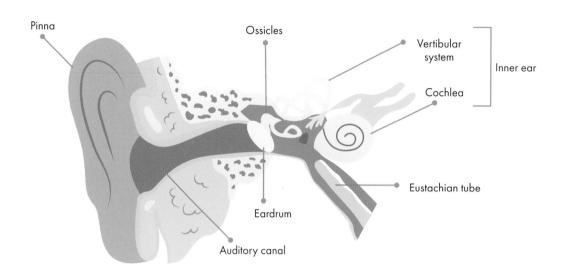

Pinna

Ossicles

Vertibular system

Inner ear

Cochlea

Eustachian tube

Eardrum

Auditory canal

As we learned, in order for a sound wave to become a sound, it must be noticed by your nervous system. Your peripheral auditory system can be roughly divided into three parts: the outer, middle, and inner ears. Your outer ear consists of your **pinnas**, or as you might call them, your ears. The pinnas help to "catch" sound waves and send them through the **auditory canal** straight toward your **eardrum**, which separates your outer and middle ears. You can help your pinnas to catch sounds by cupping your hands behind your ears as though you were listening to someone tell you a secret. Notice how you can hear better when you do this.

The sound waves get amplified further once they have entered your ear. The middle ear contains three of the smallest bones in the human body—the hammer, anvil, and stirrup. Collectively, these three bones are called the **ossicles**, and they help transmit the sound waves from the outer ear to the inner ear. When a sound wave hits the eardrum, it pushes the three ossicles that then push on the **oval window**, which separates the middle and inner ears.

Since the sound waves are moving from the air-filled outer and middle ear to the fluid-filled inner ear, some of the loudness will be lost. This is because fluid is more resistant to moving than air, so the sound waves decrease in amplitude, and therefore loudness. Think back to swimming in the pool with your sister. If she's under water and you're above water, she can't hear you unless you are yelling. Your ossicles help to increase the amplitude (loudness) of the sound waves so they can be registered in the inner ear.

Another part of the middle ear is the **eustachian tube**. This tube connects your middle ear to your throat to equalize the pressure in your middle ear to the air pressure outside your body. When there is a mismatch in pressure, you feel as though your ears are blocked up, and you may even get a sinus headache or an earache. Ever feel pressure in your ears when you dive down deep in the water or travel up high in an airplane or on a hike in the mountains? This is because the pressure outside your body is different from inside your middle ear. To equalize the pressure, you need to move air through the eustachian tube. You can do this by swallowing, chewing something like gum, yawning, or holding your nose with your mouth shut as you blow out. Don't blow out too hard, though, or you can rupture your eardrum! All of these actions lead to the opening of the eustachian tube, which will then equalize the pressure. You might even feel your ears "pop" when you do this.

The inner ear is where the auditory magic happens! It contains the **cochlea** which looks like a snail shell. The cochlea is made up of three fluid-filled canals, surrounding a bony structure containing neurons that tell the central auditory system (in your brain) that a sound has happened. When the ossicles push on the oval window, the fluid inside the inner ear moves, and it's as though a traveling wave moves throughout the cochlear canals. The movement of the fluid leads to the bending of tiny hairs on the bony structure, called **the organ of Corti**, which let the neurons know that a sound wave has occurred. These neurons then send a signal to the brain that a sound has been made.

Depending on the loudness and pitch of the sound wave, the traveling wave moves certain parts of the cochlea more than others. In fact, there is an orderly map in the cochlea based on the pitch of the incoming sound. The parts of the cochlea near the oval window (called the **base**), where the traveling wave starts, move best to high-frequency sounds (like your sister's squeal). The parts furthest away from the oval window (called the **apex**) move best to low-frequency sounds (like thunder). This is what scientists call **tonotopic organization** because the cochlea is organized by frequency, or tones of frequency. This tonotopic organization is also present in the brain areas that process sound information, including your auditory cortex.

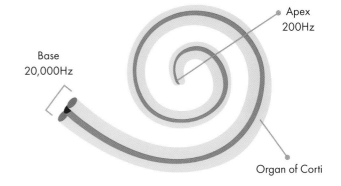

Apex
200Hz

Base
20,000Hz

Organ of Corti

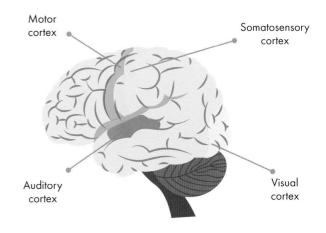

Motor cortex

Somatosensory cortex

Auditory cortex

Visual cortex

PICK YOUR BRAIN

Some people claim that you can hear the ocean if you put a seashell up to your ear. While it may sound like the ocean, it's actually just the sound of air bouncing off the inside of the shell. The shell acts as an amplifier, making the airwaves that are traveling around and in the shell louder so you can hear them. Different shaped shells make the "ocean" sound different too by exaggerating certain frequencies that are present in the air waves.

MEET DR. BRAIN

Dr. Jennifer Groh is a neuroscientist who studies how our visual and auditory senses work together to make sense of the world. Think back to when you watched a movie last. You heard the actors talking at the same time you saw their mouths moving, and so you believed their voices were coming right from where their mouths were located on the screen. But in reality, the sound was coming from a speaker, located somewhere else besides the actors' mouths. Questions like how the brain is able to combine visual and auditory information to make sense of the world is just one of the questions that Dr. Groh seeks to answer through her research.

What is a typical experiment set up like?
In a typical experiment, we present lights and sounds at either the same or different places around the lab. With this set-up, we can study how both humans and monkeys use visual and auditory information to locate items in space. We measure whether the subject can tell if the light and sound are coming from the same or different places. We also measure neural activity to see how neurons in certain brain areas respond to visual and auditory stimuli.

So how does the brain decide which sensory information to combine and how/where does it do this?
That's a good question, but the answer is a bit complicated. Few brain areas have neurons that respond only to one input. Many areas have what we call multimodal responses, meaning the neuron's activity is influenced by both sight and sound (and other sensory stimuli).The inferior colliculus is a part of the midbrain that is important for processing sound, but is also affected by sight. I've recorded neural activity from inferior colliculus neurons and have found that their neural activity changes depending on the locations of the presented visual and auditory stimuli. Sometimes the neurons fire more and sometimes less, but what is certain is that their response is affected by both the sights and the sounds. The fact that your brain seems to effortlessly combine visual and auditory information in this way leads us to conclude that these two sensory systems need to communicate with one another to make sense of the world.

Why is it important to combine visual and auditory stimuli?

Because our world is confusing, and our brain likes to make sense of what it is registering. Think about the last time you watched TV. Despite the actor's mouth moving in one place on the screen and the actor's voice originating from another place off the screen, you are still able to casually connect the voice with the mouth moving. Your brain made the connection without much effort, perhaps in part because of the multimodal neurons in certain parts of your brain.

We can also see how combining visual and auditory information is important for speech perception. Ever notice how silly things can get when you play a game of Telephone? You whisper, "My pet chicken likes to wake early in the morning" to the person next to you, who whispers it to the person next to them, and so on. By the end, the message turns out to be something like, "My vet Dr. Ken makes baked curly corn rings." Assuming everyone was trying their best to relay the correct message, this is a very difficult thing to do because lots of language sounds are ambiguous. Although you probably don't realize it, you are lip reading whenever you are listening to someone speak to help you understand what they are saying.

How do we localize sounds in our environment?

When we hear a noise, we figure out where it is coming from by comparing what it sounds like in each ear – it'll be louder in the nearer ear, for example. This is different from the visual system, where the **retina** provides a little map of where visual stimuli are in the world. Sometimes you might want to get a better look at the sound to try to figure out what made it—was that rustling a squirrel? Or a bird? To do this, you'd move your eyes to aim at the location of the sound. Such eye movements are an important part of knitting together what we see and what we hear. In my lab, we've recorded neural activity in typical auditory processing areas (like the inferior colliculus) that respond differently based on eye position. We became interested in finding out what was the earliest point in auditory processing that eye position affected neural responses. The earliest point we have looked at is actually the eardrum!

There are a couple of muscles and muscle-like cells in the middle and inner ear that control how much the eardrum vibrates. We placed a microphone in the ear canal and recorded the vibration of the eardrum. It turns out that eardrum vibrates when the eyes move. We believe that when the brain sends out a signal to the eye muscles to move the eyes, it also sends a copy of that signal to the ears to get them ready to be able to rapidly match the incoming visual and auditory information. This could be one way that the brain combines visual and auditory signals.

Dr. Jennifer Groh is a professor of psychology and neuroscience at Duke University in Durham, North Carolina.

FINAL NOTES FOR YOUR NOGGIN

Sounds are simply air waves that are picked up and registered by your nervous system. Air waves look a lot like water waves. Some are large waves, and some are small waves; the larger the wave, the louder the sound. The distance between the air waves can also vary; the closer the air waves are, the higher in pitch the sound. Sound waves are first captured by your pinna then are sent to the middle ear through the ear drum. The middle ear contains some of the tiniest bones in the body with one of the biggest jobs—they have to make the sound much bigger in order to be heard. The inner ear is filled with fluid, and sound waves cause a traveling wave to move different parts of the cochlea more than others. The moving fluid then bends tiny hairs, which signal to the neurons that a sound has occurred. That's right—neurons do the limbo and you get to hear!

CHAPTER 14

WHY IS THERE

ALWAYS ROOM

FOR DESSERT?

HAVING MY CAKE AND EATING IT TOO

You and your family have just arrived at the all-you-can-eat buffet and you're starving! You start with a large plate of salad (mostly croutons, nuts, and cheese) and then move on to your main course: mac-n-cheese, chicken nuggets, green beans, a loaded baked potato, with a side of steak. You go back for seconds of the mac-n-cheese and include some corn on the cob and a sausage. You've done it now. You're full. But wait! You didn't get dessert yet! How could you not leave room for dessert?! You decide to check out the dessert table anyway so you can see what you'd be missing out on. You spy a delicious-looking chocolate cake, some mouth-watering strawberry shortcake, and those sugar cookies that you have heard so much about. You decide that maybe you can take just a bite of each so you grab a plate. You return to the table and you end up finishing everything! How could that be? You were so full and you were sure you couldn't eat another bite of anything!

WHY DO WE EAT?

We all know that we need to eat in order to live. Food provides us with the energy to do all the things we want to do (like reading this book) and all the things we don't want to do (like cleaning the dishes after a meal). Food provides us with a range of nutrients, including vitamins, minerals, fat, protein, fiber, and carbohydrates. **Hunger** and **thirst** are the *sensations* we feel when we need to eat or drink. Actually, if you're hungry or thirsty, you're probably a little food deprived or dehydrated. This is because if you've reached this point, your body is yelling at you to get some nutrition.

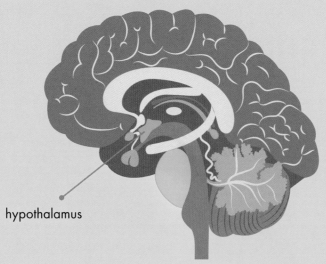

hypothalamus

The **hypothalamus** is the brain structure that helps control our feelings of hunger and thirst. It works with hormones and blood sugar levels to determine if you're hungry or if you're full in order to maintain **homeostasis**, your body's internal balance. Although your hypothalamus tells you when you're hungry or full, you can certainly ignore it! We can easily override our body's desire to eat or not. Think about all the times you decide to eat even though you're not hungry. Did you eat that slice of birthday cake because you were hungry or because you were at a birthday party and it was cake time?

From an evolutionary perspective, our bodies are always wanting to eat or drink because we evolved through a time when food used to be scarce. Early humans did not have the luxury of heading to the grocery store whenever they ran out of ice cream! They had to eat food when it was available. You can see this on the savannah, where the lion will eat his entire prey in one sitting. There are no leftovers in the wild! You can also see this at all-you-can-eat buffets, where people will graze long past the point at which they are full. People will only stop eating when it takes more energy to get up and get the food than they will gain from eating their second, or third, helping. A person might eat their third helping a bit slower than their first, but they'll keep eating until it takes too much effort to get more.

Eating is a physical necessity, but it also serves a social and cultural purpose. Sitting down to a family meal at dinner time should be a familiar ritual. We eat to get fuel for our bodies, but it also makes us feel good to share food with family and friends. In fact, satisfying your hunger and thirst might not even be the primary reason as to why you're eating dinner—you might be eating dinner just because someone said it was dinner time. The combination of physiology and behavior is partly what makes eating so very complex. Eating can bring joy, entertainment, comfort, and nourishment.

You are probably familiar with the basic taste sensations of salty, sweet, bitter, and sour. Candy companies exploit combinations of these tastes to create delectable treats like chocolate-covered pretzels and sour gummy candies. There is a fifth taste sensation called umami, Japanese for "pleasant savory taste." You can taste umami in savory foods like soup broths, meat, fish, cheese, and even some vegetables like mushrooms or tomatoes. Remember, when you say that your food *tastes good*, what you're really saying is that you like the *flavor* of the food. Flavor not only involves the taste sensations, but the smell, touch, sound, temperature, and the look of the food. Eating is a multisensory experience, as we talked about in Chapter 10. Restaurants know this! This is why they spend a lot of time "plating" their food—the chefs want to make sure your meal looks as good on the plate as it tastes in your mouth so that you will think the flavor of food is delicious!

If a food's flavor isn't what you expect, you will be less likely to eat it. Close your eyes and picture yourself eating a banana. Tastes pretty good, right? Now imagine that the banana has been colored using blue food coloring. The food coloring doesn't change the taste of the food, but since the flavor is a combination of sight and taste, you probably won't want to eat it anymore. On the other hand, if a food gives off a pleasant sensation, you may eat more of it. For example, you might be more likely to have a second piece of bread if it's freshly baked because it feels toasty and it smells amazing. Hot foods give off more **odorants** and so they smell more intense. And if something smells pleasurable, you're more likely to eat it (more on odorants in Chapter 15).

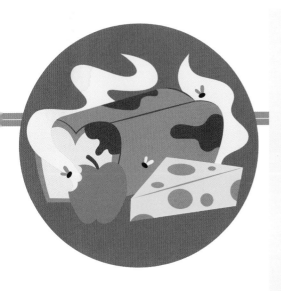

Our instinct to eat pleasant foods and not to eat unpleasant foods has an evolutionary purpose. Foods that have gone bad often have a bad smell and taste to them, not to mention that spoiled greens feel mushy! So we avoid rotten food that could make us sick. On the other hand, foods that taste pleasant are oftentimes rich in carbohydrates (bread) and sugar (candy). Our preference for eating carbs and sugars stems from our evolutionary past. As mentioned, our early ancestors evolved in a world where food was scarce, but fruits were relatively plentiful. Fruits contain a lot of sugar and carbs, which give us energy to stay alive. Early humans who developed a liking for sugar-laden fruits were much more likely to survive in the harsh elements. Our preference for sweets is a problem now because we eat too many sugary and processed foods. We can continue to eat candy and bread because they are delicious, but like most things, these should be enjoyed in moderation.

PLEASURE AND EATING

So how is it you still had room for dessert? Sometimes it has to do with the fact that you just want to eat to excess because you are not listening to your body telling you it's full. It also has to do with your perception of how pleasing you find your eating experience. As you continue to eat your favorite food, the pleasure you get from eating the food decreases, partially because the food smells less because you are becoming **adapted** to the smell. Smell adaptation can actually be a good thing, especially if it's a bad smell. Ever notice how garbage starts to stink less over time? Over time, your delicious food smells less to you, which makes the food taste less appealing. It's not that the food itself smells or tastes any less at the end compared to the beginning of the meal; it's that your brain is responding less to the taste and smell information it is receiving. So you stop eating your dinner and "feel full."

Sensory specific satiety describes the decrease in pleasure one gets from eating the same type of food. At the all-you-can-eat buffet, your body had enough of the salty, savory dinner and so you stated you were "full" and stopped eating. You should have listened to your body then, but you didn't and when the sweet dessert rolled around, you "still had room" because you hadn't been consuming that type of food during dinner. Sensory specific satiety explains, in part, why people stop eating—they get bored with their food and they stop listening to their bodies.

PICK YOUR BRAIN

You know healthy food is good for your body, but did you know that some foods are even better for your brain? For good brain health, eat these foods:

- Chocolate! Not just any chocolate, but chocolate that contains at least 80% cocoa. This is because it is rich in antioxidants, which help your cells age well.

- Berries like strawberries and blueberries are full of antioxidants as well and help keep your memory sharp.

- Fatty fish like salmon and anchovies on pizza. Fatty fish contains a lot of omega-3 fatty acids, which your brain needs to function.

- Dark leafy greens like spinach and kale are full of vitamins and minerals to keep your brain running in tip-top condition.

- Water. Dehydration leads to headaches and mood swings. No one wants those!

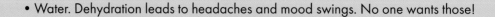

MEET DR. BRAIN

Neurobiologists like Dr. Peggy Mason study how we react to the pain we experience, as well as how we react to the pain of others. What do you do when you're in pain? And how can you make the pain not feel as bad? These are the types of questions Dr. Mason studies in her lab by using rats to study how pain is ignored when the animal is eating. She considers herself a "neuroevangelist" and enjoys teaching the public about neurobiology.

Why study rats and not humans?

We use rats in our lab because they are easier to study than humans when it comes to ingestion behaviors like eating and drinking. Humans are so varied and have a personal history that influences their taste preferences. Scientists can't control people's eating and drinking behavior, but we can control a rat's. Rats are similar to humans in so many ways. Importantly, our nervous systems work in the same way and we have very similar genetics.

What do you mean that people have a personal history that influences their taste preferences?

All people are born with innate taste preferences. For example, sweet and salty substances are innately preferred, while many bitter and sour tastes are innately avoided. However, people can change these innate preferences based on their experiences. Conditioned taste aversion is a great example of this. If someone ate something sweet, like a piece of candy, and then got sick, they might no longer like to eat that type of candy because they associate it with getting sick. On the other hand, you can teach yourself to like things you don't like. For example,

you might not like the sour taste of Greek yogurt, but since you know it is good for you, you allow yourself to learn to like its taste

.

Can you describe your typical experimental set up?

Rats are never harmed in any of our experiments. We place the rat in a cage and turn on a very bright hot light on the rat's paw. The rat will withdraw its paw in response to the discomfort it feels from the light. On some trials, we squirt a little bit of flavored water into the rat's mouth. When the rat is ingesting the flavored water, it will not withdraw its paw from the light. It seems like their tolerance to discomfort decreases while they're ingesting. We believe this is because when the brain sends a motor command to the rat's mouth to ingest, it stops all incoming sensory information, like the hot light. We call this ingestion analgesia.

Can people use the concept of ingestion analgesia to reduce our own pain perception?

Absolutely. The next time you have to get a shot at the doctor, suck on a candy or something and you should feel less pain.

Dr. Peggy Mason is a professor of neurobiology at The University of Chicago in Chicago, Illinois. Visit her on Twitter @NeuroMOOC.

FINAL NOTES FOR YOUR NOGGIN

Humans have evolved to want to eat and drink as long as there are food sources available. Food sources were scarce for our early ancestors and so, when they were available, they ate and drank until there was no more. This behavior served them well, but for us modern people who have access to plentiful gardens and grocery stores, this constant eating has led to many health problems that stem from unhealthy eating and drinking practices, like overindulgence. While the hypothalamus can help keep us in check by letting us know when we're hungry or full, it can be overridden. One reason why there's always room for dessert is because we ignore our bodies when they say we're full.

CHAPTER 15

WHY DO I REMEMBER

SEEMINGLY RANDOM THINGS

AFTER

SMELLING

SOMETHING?

Memory is a funny thing. Ever had the experience of smelling something and then being automatically transported back in time? It's as though smell is the "on" switch to your memories. Think of all the times you've walked into your grandparents' house and have smelled something cooking on the stove. All of a sudden, you're a small kid again running around their knees, tripping on your feet, grabbing the tablecloth on your way down to the floor, silverware and dishes raining down on your head. Or, when going back-to-school shopping, you smell a lead pencil and you are whisked back to your first day of first grade where you met your first best friend during recess. Why is it that smell is such a powerful reminder of past events?

OLFACTORY PATHWAY

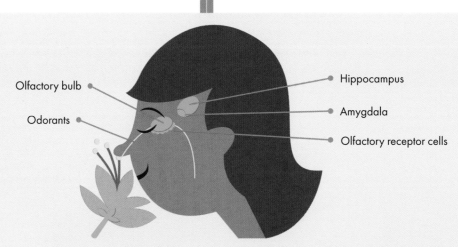

Take a deep breath through your nose. Hopefully you're not standing next to the garbage can when you do this! What do you smell? That's right—molecules! Everything we smell, from rancid meat to sweet-smelling flowers, is a combination of molecules called **odorants.**

When an odorant is inhaled, molecules travel to **olfactory receptor cells**, which are located on the roof of the nose, to the **olfactory bulbs** located on the underside of the brain. After some processing in the olfactory bulbs, the smell information goes to different places in the brain that are important for processing information about smell perception, emotional feelings, and memories. This may be why smelling something can trigger your memory of an emotional past event. When you smell your dad's homemade mac-n-cheese, the odorants travel through your nose to your olfactory bulbs. From there, information travels to an almond-shaped brain structure called the **amygdala**, which communicates with a seahorse-shaped brain structure called the **hippocampus**, to remind you of that happy day when your dad finally let you in on a family recipe. The amygdala is involved in emotional processing while the hippocampus is important for memory and, working together, they are responsible for your scent-induced memory.

PROUST EFFECT

Emotional memories that are remembered after smelling something is known as the **Proust Effect**, which refers to the vivid reliving of events from the past through a sensory stimulus, like smell or taste. The effect is named after Marcel Proust, a French writer, who wrote about the experience of a man being transported back to his youth after eating a cookie dipped in tea. The taste of the tea-soaked cookie led the man to remember his childhood Sundays spent happily with his aunt.

Scientists have discovered that when a person smells a "nostalgic odor" that brings them back to a previous positive event in their life, they feel more relaxed, less anxious, and are in a much better mood. Imagine visiting a new friend's home where you are nervous about meeting their parents. But when you walk in the front door, you smell the scent of just out-of-the-oven chocolate chip cookies. You suddenly remember all the times your favorite aunt made you cookies when you slept over on weekends growing up and you start to feel more relaxed. This is perhaps why comfort food is so comforting.

Not all scent-reminders are positive though. A single whiff of dirt might remind you of the time you fell out of a tree and had to go to the hospital for a broken arm. The Proust Effect is not only about the good times. The good and bad memories you recall when you smell something are what scientists call **autobiographical memories** because they are specific to each person. You might think of the time you fell out of a tree when you smell dirt, but someone else might remember happily gardening with their grandpa.

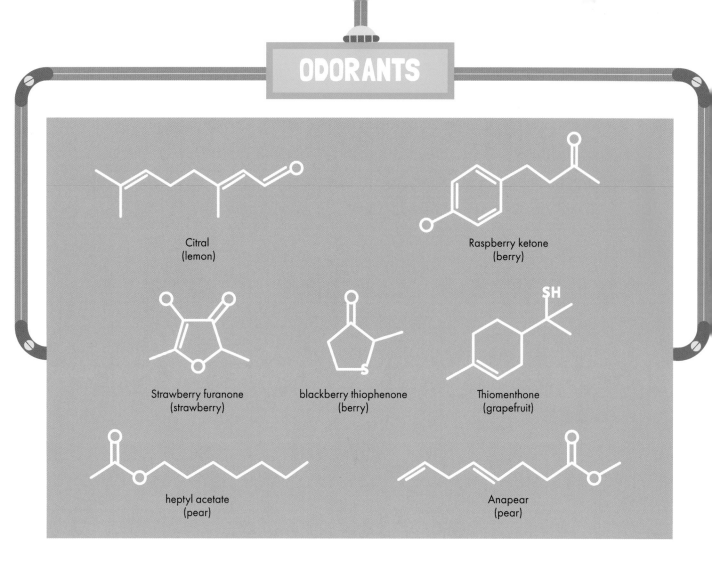

Citral
(lemon)

Raspberry ketone
(berry)

Strawberry furanone
(strawberry)

blackberry thiophenone
(berry)

SH

Thiomenthone
(grapefruit)

heptyl acetate
(pear)

Anapear
(pear)

Who doesn't love the smell of freshly cut grass on a hot summer day or the smell of winter's first snow? Some people who have to mow the lawn or shovel their driveway don't, but whether you love or hate those smells, one whiff of the air brings a mixture of molecules to your nose and memories to your mind. Mixtures of molecules form odorants like the smell of grass or snow and can be powerful memory devices. But not every molecule is an odorant. Air and carbon dioxide are good examples of mixtures of molecules that do not have a smell. In fact, odorants are added to harmful gasses like carbon dioxide so people can detect them if there is a leak. For an odorant to be smelled, it must give off vapors (like steam from hot soup) and be **hydrophobic**, meaning it must be repellent to water. Imagine last night's salmon dinner in the fridge. Fish can be quite pungent or smelly, and unless it's covered, it can stink up other food in the fridge. Uncovered butter will absorb the smell while water in a pitcher will not.

Because odorants are mixtures of molecules, the resulting number of odorants a person can smell is huge—almost one trillion (1,000,000,000,000)! While you can tell whether or not something smells, it's much harder to tell smells apart from one another. Imagine walking into a bakery. Doesn't it smell good? You can smell all the smells, but if the baker asked you to close your eyes and tell them if two smells belonged to the same kind of pastry, could you do it? Each dessert smells good, but it's hard to tell the difference between the two...at least it's hard for you and your untrained nose.

SMELL THE DIFFERENCE

Some people can learn to smell differences between substances that the average person cannot. Helen Keller was an American writer and political activist who became blind and deaf after contracting an illness when she was 19 months old. Due to her reliance on her other senses, including smell, she became very good at guessing what people did for a job (carpenter, painter, etc.) based on how they smelled. But you don't need to be blind or deaf to become an expert sniffer, you can learn to smell differences between substances. For example, people who work in the perfume industry can learn to tell the difference between a perfume that smells like roses and one that smells like cherry blossoms. Their training is like walking through a cosmetics department at a mall store getting sprayed with a bunch of perfumes over and over again and having to guess which scents are present in the perfume. Think about that the next time you get sprayed with just one perfume!

PICK YOUR BRAIN

It is possible to lose your entire sense of smell or you can just be nose-blind to certain odorants. This condition is known as complete or partial **anosmia** and can be due to infection or caused by certain medications.

Smell is considered a "mute" sense because it's hard to describe a smell without using words from your other senses. Try it yourself! How would you describe the smell of:

- butter
- chocolate cake
- the ocean
- your mom's perfume

FINAL NOTES FOR YOUR NOGGIN

Smells are made up of mixtures of molecules called odorants. When smell information enters your brain, it travels to many different areas, many of which are important for emotions and memory. Because there is such a strong link between smells and emotions, it's possible for a smell to trigger a strong emotional memory. This is called the *Proust Effect*. These memories can be good or bad and are highly personal. A smell that makes you cringe can make your friend smile. So, be careful what you sniff for!

NAME THAT SMELL!

Being able to name a smell is hard. The *tip-of-the-nose phenomenon* describes the difficulty people have with naming scents. This is different from the tip-of-the-tongue phenomenon, where people have a hard time remembering the name of something. With tip-of-the-tongue, people can make partial guesses; they might know a name starts with the letter "L" for example. There are no partial guesses in tip-of-the-nose, but people will often know what to do with the item; they can decide whether they can eat the item for example.

MATERIALS NEEDED:
- A lab partner, friend, or family member
- A blindfold
- Your Brain Journal
- Some safe-to-smell odorants (suggestions: coffee, lemon, dirt, newspaper, an orange, chocolate, etc.)

STEPS:

1 Create a simple table in your Brain Journal with columns labeled: "Name of Odorant", "Response - Odorant", and "Response - Can You Eat It".

2 Blindfold your lab partner, a sibling, or one of your parents (or have them close their eyes).

3 Wave an odorant underneath your lab partner's nose, without touching the nose.

4 Ask them to name the odorant and record that information in the table under "Response – Odorant".

5 Ask them if they think they can eat it or not and record that information in the table under "Response - Can You Eat It".

6 Repeat with other odorants.

7 Then have your friend repeat this experiment with you smelling the odorants.

WHAT HAPPENED?

Your friend should have some trouble naming the different odorants, but should have no problem letting you know whether or not the odorant is something that can be eaten. This is because the link between smells and what we call them is weak. Try this experiment on multiple people and see if any of them can smell well!

NOW WHAT?

What odorant smelled the best to you? Why?
What odorant smelled the worst to you? Why?
Why do you think our brains are able to distinguish between an odorant we can eat and one we cannot eat?

CHAPTER 16

WHY IS IT HARD

TO LEARN A

NEW LANGUAGE?

HALLO

BONJOUR

CIAO

CIAO FAMIGLIA MIA!

School's out! It's time for summer vacation! And that means you get to spend a month at your Grandma's house. You always enjoy your time there, listening to Grandma's stories, swimming in her pool, and eating her delicious spaghetti and meatballs with her homemade sauce. Plus, your cousins live right next door, and you're sure you'll spend hours biking the neighborhood and playing video games. This year will be a little different though. Your great grandma is living with your grandma. She lived most of her life in Italy and has come to live here in the United States with her children. You can't wait to get to know her too, but there is one problem. She only speaks Italian and you don't speak a word of it besides "Grazie!" which you know because your mom told you to say it often (it means "thank you"). You decide to teach yourself some Italian so that you can talk with her, but it's really hard. You keep messing up simple words like "la ragazza" (the girl) and "il ragazzo" (the boy), never mind conjugating verbs! Is it "voglio una mela" when you want an apple or is it "vuoi una mela"? Your mom and grandma can speak in Italian with your great grandma, even though they don't speak it normally. How did they get to be so good?

LEARNING A FIRST LANGUAGE

Do you remember how you learned to speak your first language? Probably not since most people start speaking when they are babies. Although there are many different languages in the world, no matter what language you learned to speak first, people go through similar phases when learning to talk. Shortly after birth, babies start to **babble** and **coo**, which is really just them practicing using their vocal cords to make the sounds present in their eventual language. For English-language learners, these sounds might include *da*, *pa*, and *ma*. Eventually, those sounds will come together to make individual words, and then they'll learn which combinations mean something (like *dada* and *mama*). The next phase is when babies start putting two words together to make phrases like *more milk, no sleep,* or *play me.* From there, babies undergo a language explosion! Babies learn new words and phrases at an amazingly fast rate— between ages one and three, they can go from knowing just a few words to over 2,000!

It was once thought that children who grew up in homes where more than one language was spoken were at a disadvantage when it came to learning to speak. The idea was that it would be confusing to have to learn two different words for the same object—is it a *ball* (English) or a *palla* (Italian) that the child wants? In fact, **bilingual** children often use words from two languages in the same sentence. This is known as **code mixing**, but this isn't actually a big deal. Many bilingual adults do the same thing, and so children are probably just mirroring what they hear, which is one way we learn any language.

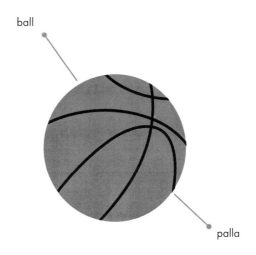

ball

palla

But don't think the children are confusing the two languages. Children do not combine words haphazardly. They can tell the difference between the languages and even two-year-olds can adjust which words they are using based on the language that is being spoken. You may not realize it, but languages differ in all sorts of ways that are not obvious at first: the types of sounds present in a language, the way these sounds are combined, and even the rhythm of speaking is different across languages. Listen to movies in other languages and compare these differences across languages. You'll see it is quite easy to distinguish between languages, even if you're not bilingual. In fact, infants as young as four months old can tell the difference between two languages!

Another myth is that children who grow up bilingual will have language development delays. The evidence for this claim is that when bilingual children are compared to their monolingual friends, they know fewer words in their common language. Let's say Rose is bilingual in Italian and English, and Madeline only knows English. If you were to count the number of words they each know in English, you might find that Madeline knows 100, while Rose only knows 80. While this may put Rose at a slight disadvantage when speaking in English, if you were to add the number of words Rose knows in Italian, 20, the total number of words known would be equal to Madeline's. In this way, Rose and Madeline's **conceptual vocabulary** is equal.

LANGUAGE AREAS IN THE BRAIN

It might seem easy to learn a language. After all, if a baby can do it, how hard can it be? But if you've ever watched a baby learn to talk, you know it is quite frustrating for them. You can experience this frustration yourself if you decide (or are required in school) to learn a second language. As you learn the second language, you should notice that understanding the language is much easier than speaking the language. These behavioral differences in the *production* (speaking) and *reception* (understanding) aspects of a language should give you a clue that these aspects of language are controlled by different areas of the brain.

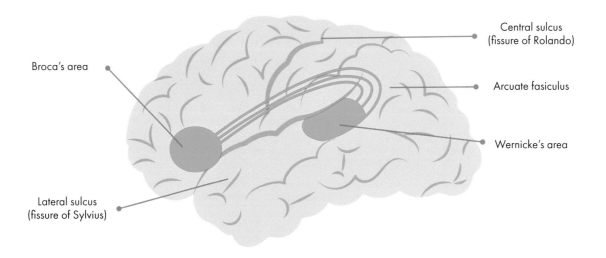

Broca's area

Central sulcus
(fissure of Rolando)

Arcuate fasiculus

Wernicke's area

Lateral sulcus
(fissure of Sylvius)

Based on research, two areas in the left side of the brain were found to be critical in language production and reception (Chapter 2). In 1861, neurosurgeon Paul Broca discovered an area in the frontal lobe near the motor cortex that was important in speech production. About 10 years later, neurologist Carl Wernicke discovered an area in the temporal lobe near the auditory cortex that was important for understanding speech. These two areas are now known as **Broca's Area** and **Wernicke's Area** and are connected to one another by a bundle of nerve fibers called the **arcuate fasiculus**. Damage to these areas leads to specific deficits in producing or understanding speech called **aphasia** (Chapter 25).

So, what happens to these brain areas when someone knows more than one language? Scientists generally agree that a person who is bilingual (speaks two languages) or **multilingual** (speaks many languages) does not have separate language systems for each language they speak. Instead, they use these same language areas, but these and other brain areas are changed by the experience of learning and speaking multiple languages.

THE BILINGUAL DIFFERENCE

By one estimate, almost one in three people in the world are bilingual. This number includes people who grew up speaking two languages, called **simultaneous bilinguals**, and those who learned a second language later on in life, **sequential bilinguals**. When a person learns a second language can influence both their behavior and brain networks. Behaviorally, simultaneous bilinguals have much better accents, have greater vocabularies, and are better at speaking each language grammatically correctly. For example, simultaneous bilinguals are better at understanding and using grammatical gender—is it *la ragazza* or *il ragazza*?—and conjugating verbs compared to sequential bilinguals.

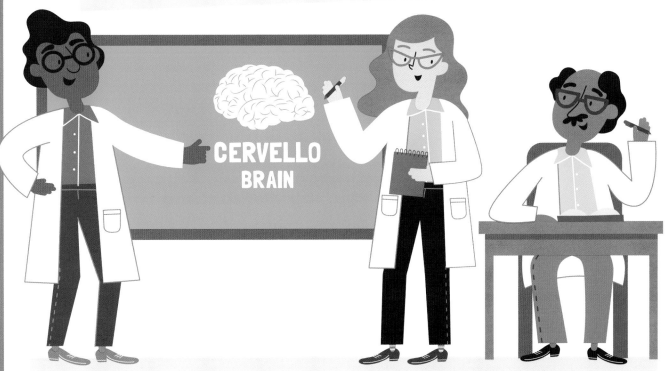

Bilinguals seem to have some advantages when it comes to cognitive functioning. These cognitive advantages may be due to the restructuring of several brain areas. Bilingual adults and children have a greater number of connections across the **corpus callosum**, the bundle of fibers that connects the left and right hemispheres (Chapter 2). This leads to more efficient communication, and less energy expended, when information moves between the two hemispheres. Bilinguals also have a greater number of neurons in their **anterior cingulate cortex (ACC)**. The ACC is important for monitoring our behavior and attentional focus, which may explain why bilinguals are faster and better at responding in tasks that require attention. Finally, bilinguals also seem to have more neurons in areas of the brain involved in working memory and impulse control, including the **inferior prefrontal cortex**. Bilinguals' behavior is a lot more flexible as they are able to stop or change an unwanted response sooner than monolinguals. They also may have some advantages when it comes to memory and attention, enhancing their ability to problem-solve.

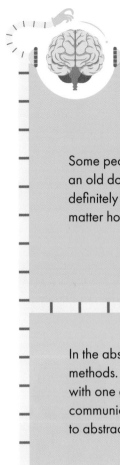

PICK YOUR BRAIN

Some people say, "You can't teach an old dog new tricks," but you can definitely learn a new language no matter how old you are.

In the absence of talking, several animals have learned to communicate using different methods. Dolphins use sounds like clicks and buzzing and pulsing noises to communicate with one another. Koko was a gorilla who used a form of American Sign Language to communicate with her human trainers. Kanzi is a bonobo who can communicate by pointing to abstract symbols that he has learned mean things like "fire," "visitor," and "yogurt."

MEET DR. BRAIN

Dr. W. Matthew Collins is a cognitive psychologist who studies how language is represented in the mind and how language is involved in memory development. What is the earliest memory that you can remember? Some people think they can remember things that happened to them when they were toddlers. However, it's more likely that they are remembering the retelling of the event from a family member and not the event itself. Dr. Collins studies why this may be the case and how knowing a language may impact your memory recall.

How does language affect memory development?

This question concerns an interesting phenomenon called childhood **amnesia**. Memory research has found that adults have almost no real memories of the events of their lives before the age of five. What they do remember is usually discussions with family members or pictures of the events—not a memory of the event itself. One theory about childhood amnesia is because children don't know as many words to describe the events they see, they don't form strong memories of those events that will last until they grow up. When we start going to school, our vocabulary grows substantially, and it is much easier to store our memories of events in the form of language, especially when we have the right words to describe those events.

How do you study this in the lab?

We test this question in the lab by examining people's memories for events that they don't have the words to describe. In our studies, we teach people the names of imaginary animals before and after they see them in an event. Then, we test their ability to remember which animals they saw in that event. When people are taught the names of the imaginary animals before they see them in an event, they are more likely to remember seeing those animals. On the other hand, when they learn the names of the animals after they see them in an event, they are less likely to remember seeing those animals in the event. Our results support the theory that memories are better when we have the words to describe what we see.

Dr. W. Matthew Collins is an associate professor of psychology at Nova Southeastern University in Fort Lauderdale, FL.

FINAL NOTES FOR YOUR NOGGIN

When first learning to speak, people go through similar phases of development, starting with cooing and babbling and ending in a word explosion. Children who grow up in homes where more than one language is spoken are not confused by the different languages. They are able to separate the two languages based on the sound and rhythm of each language. Bilingual children do not suffer from language delays and in fact have many cognitive advantages. Bilingual children and adults show more flexibility in their behavior, better memory and attention, and faster reaction times. These behavioral differences may be due to brain differences. The brains of bilinguals have more connections and more neurons in several areas of the brain, including the corpus callosum, the ACC, and the prefrontal cortex.

CHAPTER 17

WHY DO I FEEL SICK

WHEN I AM IN A

MOVING CAR OR BOAT?

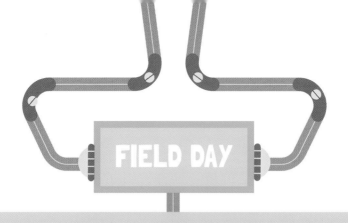

FIELD DAY

Your favorite day of the school year is finally here! Field day! That means no classes, no tests, and no lunch inside the stuffy cafeteria. This year, your school is holding field day at a local park. After attendance is taken during first period, you and your classmates walk to the front office where the school busses are waiting to take you to the park. You and your best friend have decided to sit in the back of the bus so you can talk in private, away from the wandering eyes and ears of the chaperones. You settle in the very last seat, slouch down, and start talking. Your friend wants to show you something on her phone, and as the bus takes off, you two scroll through her social media app. The bus is taking the scenic route, you suppose, as you pass through several neighborhood intersections with their speed bumps and traffic lights. After a few minutes, though, you start to feel sick. Like, really sick. You close your eyes and try to think of your happy place. Luckily, you are just about to the park. As you wait for everyone in front of you to get off the bus, you promise yourself that you're going to be sitting in the front seat of the bus from now on so you can make a speedy exit if you need to. But why did you get so sick?

You've probably learned about your five senses in school: vision, hearing, touch, taste, and smell. There is another! A sixth sense, one that is often overlooked. The **vestibular system** is the communication line that tells your brain about your head and body movements, as in which direction is up or down (gravity), and your spatial orientation (how you know when you go from sitting to standing up?). Working with your other senses, your vestibular system is the one that helps you to see clearly and to stay upright. Most of the time, the vestibular system works in the background, and we don't notice it at all as it makes small adjustments to our movements. When we do notice it, it's because we feel dizzy, sick, unbalanced, or we can't see straight. Some people report feeling **vertigo**, feeling like they are spinning or dizzy, when something is not right within the vestibular system.

The entire vestibular system is located in your inner ear, right next to the cochlea, and is about the size of a large pea! If you remember from Chapter 13, the cochlea is where our sense of hearing is registered. The word "vestibule" means *entrance*, and before scientists figured out what it did, they thought it was just the entrance to the cochlea. As it turns out, it's a lot more than an entrance! The job of the vestibular system is to detect *changes* in the types of motions described above. Have you ever noticed that when you're driving in a car for a while at a constant speed, it feels like you're not moving anymore? It's because your vestibular system has stopped responding to your motion because the motion is constant. This is a central feature of the nervous system—unless something is changing, your nervous system doesn't waste energy analyzing the incoming information.

The reason why you feel car sick is because the input from your vestibular and visual systems don't match up. When you've been in a car for a while traveling at a relatively constant speed, your vestibular system stops responding and you don't "feel" like you're moving. In contrast, your visual system is letting you know that all the things on the outside of the car – trees, houses, and other cars – are moving quickly by you, making you "feel" as though you are moving. This mismatch in the vestibular and visual sensory input is what causes you to feel ill.

Before we talk more about your vestibular system, let's look at what kind of motion your nervous system can detect. **Linear motion** lets you know whether you're moving backwards, forwards, sideways, and up or down. Next time you're in a car and backing out of a parking space, close your eyes and notice how you can sense when the car backs up, when it stops, and when it pulls forward again. That is linear motion. You can sense the second type of motion, **angular motion**, whenever you shake your head "no." In this case, your head is moving about an axis, your neck, and so the motion you feel is at an angle. Finally, there is **tilt** information, which tells you which way is up or down. It's your sense of gravity. But you don't need to be in space to appreciate gravity. Try leaning forward (but not too much!) and you will feel the pull of gravity telling you how far you've tilted toward the ground. Together, these motions allow you to sense the motion of your head, the orientation of gravity, tilt, and self-motion.

Your vestibular system is made up of three semicircular canals and two otolith organs that sense motion.

The three semicircular canals are oriented in different directions, which are called the **horizontal, superior, and posterior semicircular canals** based on their position. Think of the semicircular canals as three donuts, each with a bite taken out of them and connected where they were bitten. These canals are filled with liquid, and when you move, the liquid moves and bends tiny hairs inside the canals. It is the bending of these tiny hairs that lets you know you're moving (much like how it's the bending of the tiny hairs in your cochlea that allow you to hear).

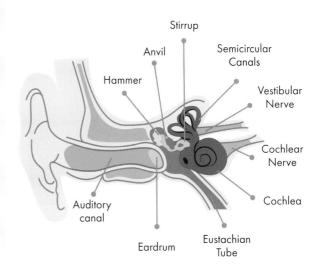

The two **otolith organs** are located between the semicircular canals and the cochlea (called the **utricle and saccule**) There word "otolith" is Greek for "ear stones." Your otoliths actually contain small stones made of calcium carbonate crystals. These tiny crystals move through a gelatinous structure and cause the tiny hairs in the structure to bend. The bending of the hairs is what lets you know whether you're moving and the position of your body!

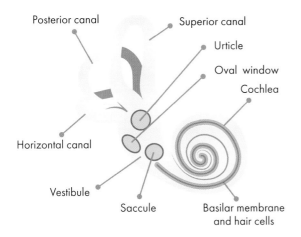

You can now see how the vestibular system, with its three semicircular canals and two otolith organs, senses motion. As you move about, the motion causes the fluid or the stones to move, leading to the tiny hairs bending and you sensing the movement. If you keep on moving at a constant rate, eventually the fluid and stones will stop moving, the tiny hairs will no longer bend, and you will no longer sense motion. That is why you stop feeling like you're moving if you're in a car or plane that is traveling at a constant speed. If the car you're in stops suddenly or if you feel turbulence in the plane, your vestibular system will come back online, and you'll feel the change in movement.

On the other hand, if you stop suddenly, the fluid and stones will keep moving for a while due to inertia. This is why you feel dizzy after you spin around real fast and then stop. You stopped, but your vestibular system is still telling you that you are moving. This mismatch of vestibular and visual information is what makes you feel dizzy.

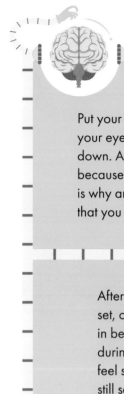

Put your vestibular system to the test! Spin around in an office chair at a constant rate with your eyes closed. After about 20 seconds, you'll start to feel as though you're slowing down. After about a minute, you'll feel as though you've stopped completely. This is because your vestibular system has stopped responding to the constant rate of motion. This is why amusement park rides have to keep changing the direction and speed of the rides so that you feel as though you're still moving about—and perhaps even get sick in the process!

After a day spent at the skating rink, bobbling in the surf at the beach, or on a swing set, one can often feel a kind of ghost of those regular, alternating movements while in bed that night. This is because your body learned to adapt to the constant motion during the day by sending out an equal and opposite motion signal to make you feel steady. Now that you're not moving and there's no real movement, your body is still sending out that equal and opposite signal so you still feel like you're moving.

MEET DR. BRAIN

Dr. Joshua Bassett is a neuroscientist who studies the vestibular system in rats. Using very tiny wires, he studies the brain activity of these small animals as they are moving throughout their environment. He records from a group of cells called head direction cells, which are active when the animal's head is pointed in a specific direction. We asked him what these head direction cells were good for and how he studies them.

What is a typical experimental set up like?
I study how rats are able to find their way through their environment by measuring both behavioral and neural data. In particular, I study a group of neurons called head direction cells in the rat brain. These cells are located in the thalamus and are connected to areas that are related to memory, like the hippocampus. These cells are sort of like the rat's compass system and help the rat with its sense of direction. In a typical experiment, I place a rat in a maze and then watch as it tries to find its way out. I also record from these head direction cells while the rat is trying to figure out which direction to move in. Head direction cells fire best when the rat's head is facing a particular direction, with different cells preferring different directions.

What is the function of head direction cells?

It turns out that these compass-like head direction cells are able to create an internal map of the rat's surroundings. But the map is not created using the same signals that a real compass uses. A compass uses a geomagnetic signal created by the Earth's North and South poles to orient itself. Head direction cells seem to orient the rat relative to big landmarks in their environment, like a large X that we place on one wall of the lab. When the rat can't see the landmark, they appear to be confused and wander around in much the same way that you might be confused if you came up from an underground train station and got turned around. The head direction cells seem to be confused as well and do not give a clear pattern of neural firing indicating which direction is which. But once the rat finds the landmark, they begin to act like they know where they are going, and their head direction cells start firing more predictably.

How does research here on Earth help astronauts in space?

Astronauts suffer profound illusions that disorient them, and this can have severe consequences for their well-being and for their mission objectives. One disorienting illusion is when astronauts feel as though they have flipped upside down because their perception of what's up and what's down has changed. This is especially disorienting if the astronaut is on a space walk because this upside-down illusion can translate into the feeling of falling toward the Earth, which can be terrifying. Another illusion is when they have trouble with their spatial memory after being in a small confined space for so long. They essentially find it difficult to remember where things are located and can confuse their left and right.

In one experiment, we sent rats up in a plane that simulated zero gravity (0G) to study how head direction cells responded. Although true 0G doesn't happen until you're well out of the Earth's atmosphere, you can achieve microgravity in a plane that follows a parabolic flight trajectory. On a parabolic flight trajectory, the plane travels almost straight upwards and then downwards. On the downward portion of the trajectory, it can feel as though you're in 0G. On such a flight, we measured how head direction cell firing changed as the animal entered microgravity. We found that unlike in regular gravity (in 1G), head direction cells did not fire less when the rat was upside down. Instead, we found that the cells started to respond to a location that was 180° from where it responded in 1G. This may partially explain the upside-down illusions that astronauts sometimes feel when they're in space.

Dr. Joshua Bassett is a postdoctoral fellow at the University College London in London, England.

FINAL NOTES FOR YOUR NOGGIN

Your vestibular system is designed to detect self-motion and spatial orientation. It is made up of 3 fluid-filled semicircular canals and two stone-filled otolith organs. When you are moving, the fluid and stones move and bend tiny hairs, which signal to your nervous system that you are moving and your position in space is, too. If you continue to move at a constant rate, the fluid and stones will stop moving, leading to no more bending of the tiny hairs and a feeling as though you are not moving. This can happen when you're in a car that is traveling at a roughly constant speed down the interstate. But your visual system still registers your motion because all the houses, trees, and other cars are moving quickly past you. This mismatch between what your vestibular and visual systems are telling you can make you start to feel dizzy, or even sick. One thing that may help with motion sickness is to look far off into the horizon where objects don't appear to be moving as much. In this way, you can start to match you visual and vestibular inputs better, and feel better.

CHAPTER 18

WHY CAN'T WE TICKLE OURSELVES?

You're pretty excited for this vacation! Your four-year-old cousin is coming to visit, and you haven't seen her since last year. That's when you visited her, and you had so much fun with her. You played in the snow, ate ice cream, and watched cartoons for most of the visit. You also played "Monster" a lot—a game where you chased her around the house with her screaming, "Come get me!" When you did finally get her, you tickled her, and she laughed her infectious laugh. Then you let her go, and the game started all over again. It seems as though she could play that game forever! Truth be told, you were getting a little tired of chasing her all around and you finally said, "The monster is sleeping and can't chase you anymore." She stuck out her bottom lip and pouted. When that didn't "wake" the monster, she begged, "Please monster! Please chase me! Please tickle me!" You told her you just needed a few minutes to rest and suggested that she chase and tickle herself. That worked for a bit. She ran around the kitchen screaming and then plopped herself down next to you on the couch and started tickling herself. She tried to laugh, but it wasn't the same. She was tickling herself the same way you had been doing (by squeezing her belly), but she claimed she just didn't *feel* tickled. You started to wonder why people can't tickle themselves and then reached over and grabbed her belly for a good tickle and laugh.

The skin has many useful purposes, including keeping all your guts in your body! And your skin does much more than that, it protects you from the UV rays of sunlight that can damage your cells, is the first line of defense against infection, and it helps to direct blood flow to all parts of your body. Skin is also the way we get information about the world—how cold or hot it is, how windy the day is, and whether or not something (or someone) is touching us.

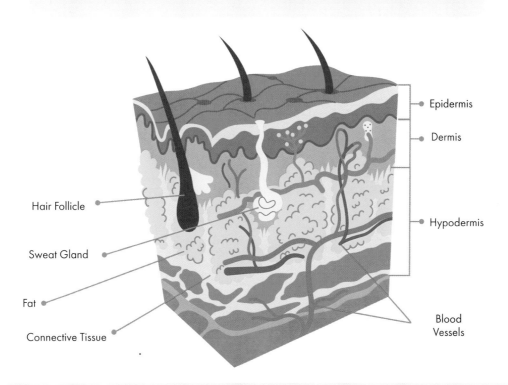

Epidermis

Dermis

Hypodermis

Hair Follicle

Sweat Gland

Fat

Connective Tissue

Blood Vessels

Your skin is a layered structure and can be roughly divided into three parts. The **epidermis** is the layer of skin that you see and is what creates your skin tone. Below that is the **dermis** layer where hair follicles and sweat glands are. Finally, there is the **hypodermis** layer ("hypo" means under in Greek—so this is the layer that is *under* the dermis). This layer contains fat and connective tissue. Throughout your skin, there are specialized receptors for touch called mechanoreceptors. Different mechanoreceptors respond to different kinds of touch information, including tickling sensations. When these mechanoreceptors are activated by tickling, they send a message to the **somatosensory cortex** (Chapter 11) that something (or someone) has touched you. Notice that the message sent to the brain is that "you've been touched", not "you've been tickled." In order for you to feel tickled, a lot more needs to happen.

It is very easy to tell the difference between whether you are touching yourself or if someone is touching you. Imagine being touched on the upper part of your knee, a place many people find ticklish. If you reach down to touch your knee, your brain is expecting the touch. There is no surprise there. If someone else touches your knee, it is somewhat unexpected. This is part of why tickling can be fun—it is unexpected! Scientists have discovered that the cerebellum (Chapter 1) may be the part of the brain that prevents us from tickling ourselves. The cerebellum is located at the base of your brain, and one of its many functions is to monitor our movements. It can tell the difference between touches that are expected and those that are unexpected. So, when you reach down to touch your knee, your cerebellum tells you that it was *you* touching you. You're not surprised because you knew your knee was going to be touched. When someone else touches your knee, your cerebellum did not know that was going to happen, so it's unexpected, and you might laugh because it felt ticklish to you. But it takes more than an unexpected touch to make you feel ticklish.

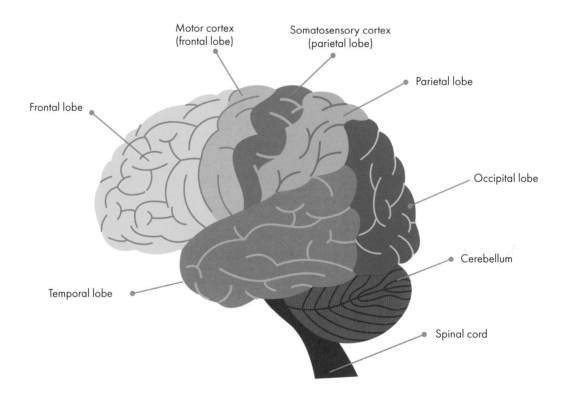

Motor cortex (frontal lobe)

Somatosensory cortex (parietal lobe)

Parietal lobe

Frontal lobe

Occipital lobe

Cerebellum

Temporal lobe

Spinal cord

As you well know, not all unexpected touches make you ticklish. If someone accidentally bumped your knee in the same place you usually find ticklish, you might not find that funny. Another part of the fun of being tickled has to do with the circumstances under which you are being touched. If you were playing around fighting with your cousin and they accidentally hit your knee, you might find it ticklish. But if you were in an angry fight with the person who hit your knee, you might feel pain (Chapter 8). So, in order for the unexpected touch to be felt as a tickle, you have to find the situation you are in fun. The brain encodes messages in context!

The area of the brain that processes all kinds of pleasant information is called the **anterior cingulate cortex** (ACC). It is located in the front of the brain and is involved in all sorts of higher order cognitive functions like feeling empathy, making decisions, and controlling one's impulses. The ACC is also the brain area that becomes activated whenever you smell something pleasant, eat something pleasant, touch something pleasant, or if you're in a pleasant situation. It's one of the brain areas that lets you know when things are good or bad.

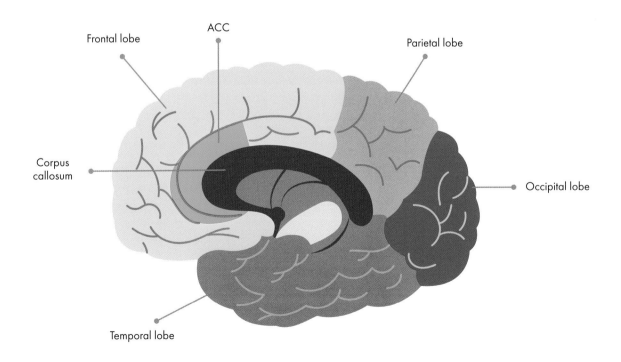

Frontal lobe

ACC

Parietal lobe

Corpus callosum

Occipital lobe

Temporal lobe

PICK YOUR BRAIN

Your skin is the largest organ in your body. It is also the heaviest, weighing in at around eight pounds in a full-grown adult. That same adult has a skin surface area of about 22 square feet! To give you an idea of how big that is, look at any standard door, which has a surface area of about 21 square feet. Think about that next time you're hiding behind a door!

FINAL NOTES FOR YOUR NOGGIN

Tickling is a team sport—many brain areas have to work together to make it happen. Feeling ticklish is so much more than just feeling touched. In order for you to feel tickled, the touch has to be both unexpected and pleasant. Your cerebellum lets you know the touch was unexpected—did you touch you or did someone else touch you? Your ACC lets you know if the situation you are in is a good one. You'll only find a touch ticklish if you're in a fun situation. Your cousin's ACC was telling her she was having fun, but her cerebellum was telling her that she was the one doing the tickling, so the touching wasn't unexpected. Looks like the "Monster" is just going to have to wake up and tickle her until you both become exhausted with laughter!

BUILD A TICKLE MACHINE

Although it's hard to tickle yourself, it's not impossible. The key is to touch yourself very lightly in a somewhat unpredictable way.

MATERIALS NEEDED:

- A 12-inch ruler
- A long feather
- Tape (something to attach feather to ruler)
- Your bare foot
- Your Brain Journal

STEPS:

1 Attach the feather to one end of the ruler.

2 Get comfortable and in a good mood.

3 Cross your legs at the knee so that the sole of one of your bare feet is exposed.

4 Gently pass the feather over your foot (the bottom, the top, the toes).

5 Keep track of which sensations make you feel most ticklish in your Brain Journal.

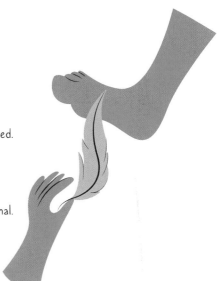

WHAT HAPPENED?

If you were lucky, you just tickled yourself! The tickle and the resulting laughter might not have been as extreme as it would be if someone else were tickling you, but you probably felt a tickle or two as you moved the feather across your foot. By having the feather do the actual touching, your cerebellum is not quite sure where you are going to be touched so it can be unexpected. By doing this in a good mood, your ACC is letting you know that this is going to be fun!

NOW WHAT?

Ask a friend to tickle your foot with the feather and see if you feel more ticklish. Try using a different object, something harder, like the other end of the ruler. Are you more or less ticklish the more pressure you put on your foot? See if you can find a combination of places on the foot and pressure of the feather that makes you laugh when your friend is controlling the feather, but not when you are holding it.

CHAPTER 19

WHY DOES

MY HEART RACE

WHEN I'M SCARED?

It's summer! And that means it's time for a visit to your favorite amusement park! You can already taste the cotton candy and caramel apples, and don't even get you started about the fried dough! You can feel the powdered sugar on your lips and nose and chin as you think about taking that first hot bite. You also really can't wait for the carnival games! You set up a couple of two liter bottles and fashioned some rings out of pipe cleaners and have been practicing your ring toss all year long. You even made a makeshift water game by cutting a hole in the shower curtain and using your water gun to shoot into the "clown's" mouth. Your balloon *will* fill the fastest this summer and you'll finally win that huge purple snake they keep tucked up in the corner reserved for the big winners. But your excitement starts to wane when you think about the rides. You don't mind the bumper cars, the scrambler, or even the roller coasters, but the ferris wheel terrifies you! You have never been a big fan of heights. Just thinking about slowly circling up and over the entire carnival makes you queasy. Your heart starts to race, and your palms get sweaty. Even the thought of being up there makes you nervous. You wonder why you react this way and wish there were a way that you could learn to calm down.

Everyone is afraid of something. What are you afraid of? From an evolutionary perspective, fears are extremely helpful. They let us know when we might be in danger so we can react to the danger in order to survive. For example, let's say you're walking home from a friend's house and it's late at night. You have the option of walking past a house with a vicious dog or crossing the street to avoid the terror. As you are approaching the house, your heart starts racing and your breathing becomes quicker and shallower. These responses let you know that something dangerous is coming up. This is the survival instinct you need in order to motivate yourself to cross the street so you can avoid being bitten by that dog.

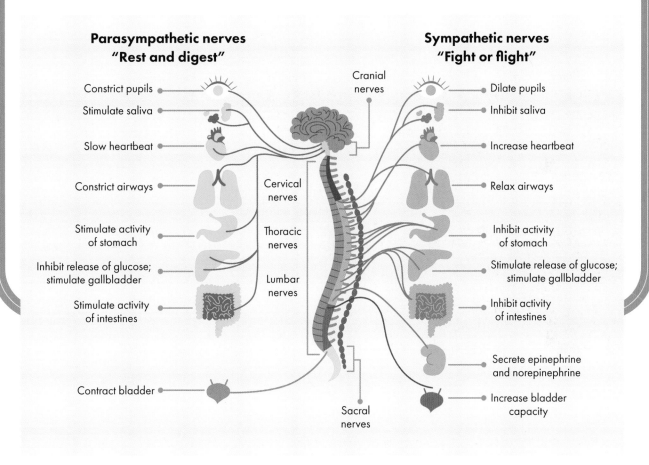

When we are scared, our body prepares us to react to the object or situation that is making us fearful. As mentioned in Chapter 3, your nervous system can be roughly divided into two parts—your central nervous system, which includes your brain and spinal cord, and your peripheral nervous system, which includes all your other nerves. Your **sympathetic nervous system** and **parasympathetic nervous system** are parts of your peripheral nervous system. These two systems help to keep your body in **homeostasis**, or in a state of equilibrium. If you are just chilling at home sketching on your notepad or watching your favorite TV show, your parasympathetic nervous system is keeping your body going by allowing you to "rest and digest." When this system is activated, your heart rate slows, your body digests whatever food you've eaten, and you might even feel the urge to visit the bathroom. On the other hand, if you are in a scary situation, your sympathetic nervous system is helping your body get ready to "fight or flight." Your heart rate speeds up, your body stops doing non-essential tasks like digesting your lunch, and hormones are released that prepare you to react quickly.

WHERE DO FEARS COME FROM?

Why is it that some people are afraid of heights and some people are not? It may be that some fears have a genetic basis. This is not to say that your DNA makes you fearful of things, but rather your DNA may help explain your reaction to certain fearful situations. People vary in their reactions to fearful situations and that variability might partly explain whether someone might develop a **phobia**. A phobia is an extreme fear of something, someone, or some situation. A person with a phobia isn't just scared, they're terrified, and this phobia interferes with their everyday life. They might avoid a situation, place, or person, and if they're confronted with the fearful thing (or even if they think about it), their heart starts to race, their breathing gets faster, and they may start to shake. In other words, their "fight or flight" response goes into full gear.

While there may be some genetic basis for fear, most scientists agree that a person's prior experience explains why certain fears develop. For example, if you almost drowned as a child, you might develop a fear of the water. It's not usually that easy though to pinpoint the exact moment when a fear might have developed. Think about something that scares you. Do you remember when you first started being afraid about whatever it is that scares you? Most people can't do this, but just because you don't remember how a fear started, doesn't mean the experiences aren't very real.

Most scientists believe we are born with few innate fears. One fear we're born with is that of extremely loud noises, which is why newborns get startled and may even start to cry if a loud noise happens around them (do *not* try this one at home!). Since we're born with so few fears, the majority of our fears are developed based on our life experiences. The idea that fears are learned leads many to believe that fears can be unlearned. Cognitive behavioral therapists are trained psychologists who help people change the way they think and act when they encounter something that makes them scared. They do this by using a variety of **exposure therapies**, which expose their clients to the things that make them scared while the therapists talk them through it.

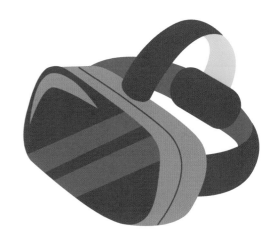

For example, if you're afraid of heights, a cognitive behavioral therapist will help you to relax when you think of being up high. They might even get you to wear virtual reality goggles so you can "practice" being up high on a ferris wheel (more on 3D vision in Chapter 7) while you are safe and sound in the therapist's office. As you safely go around and around the VR ferris wheel, the therapist will talk to you and help you learn how to relax and will help you change the way you think about being up high. Most people with a fear of heights say they're afraid because they think they'll fall down. A therapist can help the person reframe their way of thinking so that instead of thinking that they will fall down whenever they are up high, they instead think about how unlikely it is that they will fall. This kind of therapy can take some time, but many people feel better able to deal with their phobia after one brief session.

Phobias are extreme fears that some people have when faced with particular objects or situations. Some specific phobias include:

- Arachnophobia – fear of spiders
- Acrophobia – fear of heights
- Aerophobia – fear of flying
- Social phobia – fear of social situations
- Cynophobia – fear of dogs
- Trypanophobia – fear of needles

- Alektorophobia – fear of chickens
- Claustrophobia – fear of enclosed spaces
- Xenophobia – fear of the unknown
- Nyctophobia – fear of the dark
- Astrophobia – fear of thunder and lightning
- Mysophobia – fear of germs

MEET DR. BRAIN

Dr. Jens Foell is a cognitive neuroscientist who uses neuroimaging like MRI and EEG (see Chapter 5) to understand the connection between a person's brain activity and their behavior, perception, or personality. His research has focused on fear processing in people, including those high in psychopathy (lack of empathy, impulsiveness, callousness, and a stunted fear response), as well as how brain pathways are rewired after traumatic injury.

How do you study fear in the lab?

Being a "neuroguy," I am interested in what the body and brain is doing when people are experiencing fear. In the lab, we can induce a fear response in people a couple of different ways. One way is to startle them with a loud sound or a mild electric shock while we measure their eye muscle activity. When people are startled, they blink their eyes. We find that the intensity of the eye blink is related to how fearful people are feeling—stronger eye blink responses mean the person is feeling more fear. We have also noticed that people with intense startle responses seem to be more susceptible to developing certain fear disorders like phobias. Another way we study people's fear responses is to compare their physiological responses to both fearful images, such as a vicious dog baring his teeth, and neutral images, such as a field of flowers. While viewing images, we measure people's heart rates and skin conductance levels, or how sweaty their palms are. Fearful images lead to higher heart rates and sweatier palms. We combine all these physiological measures—muscle activity, heart rate, skin conductance levels—and compare them to questionnaires where we ask people just how much fear they're feeling. Using multiple measures of fear processing gives us more confidence in our results.

How might you study brain activity while someone is scared?

We usually have people look at fearful or neutral images one at a time on a computer screen. We have them either just look at them or rate them in some way so we can make sure they're paying attention. In an EEG recording of brain activity, we can measure just how quickly or slowly people react to fearful stimulus. In an fMRI experiment, we can determine where in the brain fear is being processed.

So where in the brain is fear processed?

Most people find that the amygdala is involved in fear processing. The more active the amygdala is, the more fear people report. On the other hand, in people whose amygdala has been damaged either due to a stroke or some other traumatic brain injury, they do not feel fear at all. You might think that a person with a large amygdala has a bigger fear response, but in fact, there doesn't seem to be a strong relationship between the size of the amygdala and a person's fear response.

What is psychopathy and how is the fear response different in people who are high in psychopathy?

Psychopathy is a disorder that is characterized by a combination of traits. Some of these traits include having a lack of empathy for other people, being impulsive, feeling callous, and even having a stunted fear response. In the lab, we can induce a fear in people using something called fear conditioning. Similar to classical conditioning (see Chapter 4), we show people pictures of people they don't know and sometimes we pair the picture with a mild electric shock. People not on the psychopathy spectrum start to get anxious when they see a face that has previously been paired with a shock. People high in psychopathy do not get anxious. Fear conditioning does not work in these people. This tells us that these people do not learn to fear the way that others do.

Dr. Jens Foell is an Associate in Research in the Clinical Psychology Department at Florida State University in Tallahassee, FL.

FINAL NOTES FOR YOUR NOGGIN

Everybody is afraid of something. Some fears are actually healthy because they ensure that we remain safe. Your sympathetic and parasympathetic nervous systems help to keep your body in homeostasis (equilibrium) between cycles of "rest and digest" and "fight or flight." When we are afraid of something, our "fight or flight" system kicks into high gear and our hearts start racing, our breathing becomes shallow, and our palms become sweaty. Phobias are extreme fears of something, someone, or some situation. Cognitive behavioral therapists can help people with their fear responses through exposure therapy where they expose the person with the fear in a safe environment, sometimes with the aid of VR goggles.

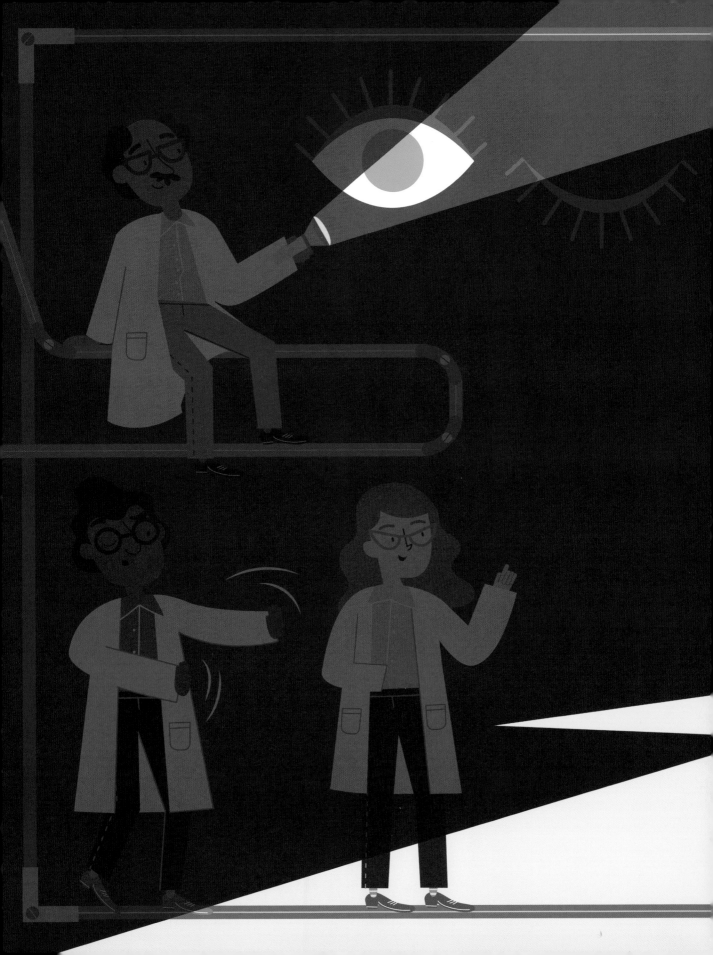

CHAPTER 20

WHEN

I SHUT OFF THE LIGHTS IN MY ROOM, I CAN'T SEE ANYTHING,

BUT AFTER A FEW MINUTES, I CAN.

WHY?

I CAN'T SEE YOU, OR CAN I?

Imagine that it's the middle of the night and you wake up thirsty. You use the little bit of light that is filtering in through your bedroom curtains to navigate your way to the bathroom outside your room. You turn on the light, fill up your cup with water, drink, and head back to bed after shutting off the bathroom lights. Suddenly, you're plunged into complete darkness. The light from the window barely illuminates anything in your room, and you trip over your clothes that you left on the floor instead of putting in the hamper. You finally lay down and wonder how your room got so dark so fast. After about 10 minutes, you notice light is starting to illuminate the objects in your room again, including those dirty clothes. What just happened?

LIGHT

We all know we need light to see, but what exactly is light? Light is the movement of very tiny particles that carry energy, called **photons**. The photons move at different rates of speed and are carried along by a wave of energy. This energy is detected by cells in our eyes. We see a rainbow of colors because different colors are produced from different wavelengths of light energy. When a ray of light shines on an object, there are several things that can happen to it. The light can be *absorbed* by the object. Dark objects absorb more light energy than white objects and convert it into heat. This is why dark shirts make you feel hot in the bright sun of summer. The light can be *refracted*, or bent, especially if it is traveling through different mediums. When you go swimming, if you look at your foot under the water, it appears to be distorted. The reason is the light rays bend as they go from water to air. Light can also be *reflected* from an object. Look in a mirror and you'll see your reflection. The light reflects off your face to the mirror and then reflects back into your eyes. Cells on the back of your eyes register the light and send a signal to the brain that you're looking at something.

Parts of the human eye

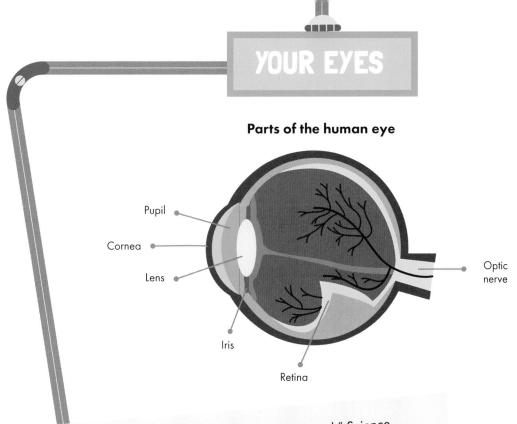

Pupil

Cornea

Lens

Optic nerve

Iris

Retina

Some people say that the "eyes are the window to the soul." Science can't prove that, but they are indeed a window into your nervous system. Eyes are the only place in your body that a doctor can look directly at your nervous system, with the use of an **ophthalmoscope**. The ophthalmoscope shines a light in your eye, which then reflects off the back of the eye into a tiny hole in the ophthalmoscope. The doctor is able to see the different structures of your eye to make sure they are healthy. There is a structure that covers the entire back surface of your eye and is the part that contains the cells that convert light energy into the language of the nervous system, action potentials (see Chapter 1). This structure is called the **retina**.

Light first enters your eye through a tiny hole called the **pupil**. This is the black circle right in the middle of the colored portion of your eye, the **iris**. The iris is what holds the color and shine in your eyes, but its purpose is more than to just look pretty! It is what controls the diameter and size of the pupil. The size of your pupil determines how much light gets in your eyes and can be changed by how dark or bright it is outside. When you're in your darkened bedroom, the pupil gets bigger, allowing for more light to enter your eye. When you turn on the light when you get your drink from the bathroom faucet in the middle of the night, the pupil gets smaller. Try looking in a mirror and turning the bathroom lights on and off; you will see your pupil changing size as the light changes! This is important because your eye requires just the right amount of light to see well. If too much or too little light is let in, then you won't see clearly.

Once light enters the eye, it passes through the **lens**, which helps to focus the light on the back of your eye. Your eye's lens works just like glasses in this way. As light passes through a lens, whether in your eyes or in your glasses, it focuses the light on one place at the back of your eye on the **retina**. The retina contains cells that are sensitive to light called **photoreceptors**. **Cones** and **rods** are two types of photoreceptors, named so because of their shapes. Cones are shorter than rods and have more pointed tops. Each photoreceptor contains light-sensitive proteins (discs) that absorb light. The absorption of light by the photoreceptors starts a process that ends with the brain getting a signal that light was present. Different photoreceptors respond best to different types of light. The cones respond best to bright lights and the rods respond best to dim lights. So, you're using mostly your cones to see at the beach and mostly your rods to see when you're in a darkened room at night.

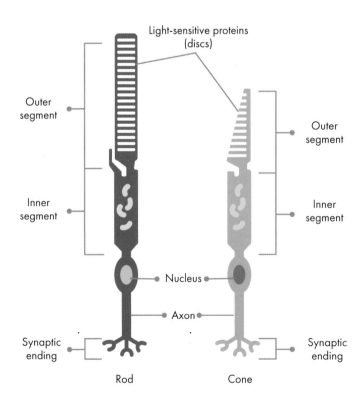

Light-sensitive proteins (discs)

Outer segment

Inner segment

Synaptic ending

Outer segment

Inner segment

Synaptic ending

Nucleus

Axon

Rod

Cone

No matter what time of day, once the light hits the photoreceptors, it takes time for them to be ready to respond to another ray of light because the light-sensitive proteins are all used up. In light, the proteins are used and regenerated at the same rate. In darkness, the proteins have a chance to build up. The more proteins you have, the more sensitive you are to light, meaning it takes a lot less light to send a signal to the brain. Dim lights seem bright and bright lights seem painful! This is why your eyes hurt when the lights are turned on after being in a dark movie theater.

This process of becoming sensitive to light after spending time in the dark is called **dark adaptation**. When you're in a dark room, the proteins have a chance to build up, so you're able to see in the dim light of your bedroom. Once you turn on the light in the bathroom, not only does the bright light make you squint because the light hurts your eyes, but all the light-sensitive proteins are being used up. When you go back to your dim room, it takes a while to see in the "dark" since it takes time for the proteins to build up again.

PICK YOUR BRAIN

When you look at something like a tree, the light bounces off the tree and into your eye through the pupil. The light is then filtered through the lens, which turns the tree upside down on your retina. Your brain then flips the image back over so that you see the world right-side up. All lenses work this way. Try it for yourself by looking through a magnifying glass.

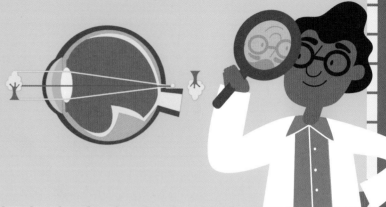

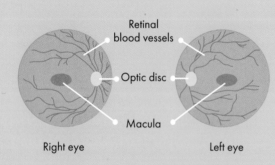

Retinal blood vessels

Optic disc

Macula

Right eye   Left eye

The diagram shows what the insides of your eyes look like to your doctor. The blood vessels are the squiggly lines, and you can notice that they all come together in the bright region of each image called the **optic disc**. This is the area of the eye where all the blood vessels and nerves enter and leave the eye. The dark region in the center of each image is called the **macula**, which is part of the **retina**. When you look at something, the light is focused through your lens to the macula. So, the next time you want to say, "I've got my eyes on you," you instead should say, "I've got my maculas on you!"

FINAL NOTES FOR YOUR NOGGIN

In order to see, light passes through your pupil and lens and eventually becomes focused on your retina. The retina contains photoreceptors, the cells that contain light-sensitive proteins and send a signal to the brain that you are looking at something. When light hits one of your photoreceptors, the proteins are used up and it takes time for the proteins to be ready to be used again. When you wake up to get a drink of water and turn on the lights, the light-sensitive proteins get used up. So, when you first get into a "darkened" room, you do not notice the little bit of light that may be present. However, over time, the light-sensitive proteins are replenished, and you are able to see in a "dark" room. This process of dark adaptation is why your dark bedroom starts to look a lot brighter as time goes on.

ACT LIKE A PIRATE

Being able to "see" in the dark is one advantage of wearing an eye patch, and it would be especially handy if you were a pirate on a ship. When you are out on deck in the bright sun, your uncovered eye would allow you to see normally. Since your covered eye is not using up the light-sensitive proteins, that eye is becoming more sensitive to dim lighting conditions, like what you find when you go below deck. So, when you go to your bunk down below, you would be able to see in the "dark" if you use the eye that was covered because it's been dark adapted. Let's try this out ourselves!

MATERIALS NEEDED:
- Eye patch, hands, or something else you can use to cover one eye
- A timer
- Your Brain Journal

STEPS:

1 Place the patch over one of your eyes for 15 minutes. With only one eye, your vision and depth perception may be poor, but feel free to do whatever you normally would do while your eye is covered.

2 After the 15 minutes are up, carefully remove the patch and note how the light levels are different when you look around with each eye by opening and closing one eye at a time. The covered eye should make everything look brighter.

3 Record what you observe about the difference in light levels in your Brain Journal. You could use one page to write or draw what you see with the eye you had covered and one page to write or draw what you see with the uncovered eye. What are the light levels like when you look around with the eye you had covered? What are they like for the eye you left uncovered?

WHAT HAPPENED?

The covered eye made everything look brighter because while you were covering the eye, the light-sensitive proteins in that eye had time to build up. Since there are more light-sensitive proteins to absorb the light when you uncover your eye, the light appears brighter. You can test this by going into a darkened room and noticing how you can see in the dark with the covered eye and not the other eye.

NOW WHAT?

You just tested what the world would look like when you covered one eye for 15 minutes. Try the experiment several more times, but vary the amount of time you keep the eye covered to see how long it would take to become fully dark adapted. Do this in a darkened—but not dark—room, and to test how well your eye is dark adapted, try reading from a book every time you take the eye patch off with the now uncovered eye. Record your observations in your Brain Journal.

CHAPTER 21

WHY DOES GRANDMA KEEP FORGETTING THINGS?

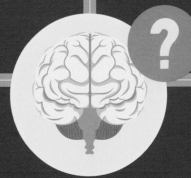

It's Thanksgiving, so that means you get to hang out with your whole extended family at your grandparents' house. You can't wait to talk with Aunt Tara about your newest obsession, jewelry making. You've recently gotten into bead crafts, and you're hoping to get some tips because she's been doing it for years. You're even going to bring your bike over so that you and your cousins can bike around the neighborhood after lunch. When you arrive, Grandpa greets you at the door and gives a sad look to your Mom saying, "Today's been a rough day so far, but I'm hoping it gets better. She usually does better after lunch." You're not sure who he is talking about until he looks at you and says, "Hey honey. Grandma isn't quite herself right now, don't feel bad if she acts strange, ok?" You say ok and then move inside the house to say hi to Grandma. You find her sitting in the living room, looking out the window. "Hi, Grandma!" you say as you walk toward her, hands reaching out for a hug. She turns her head to look at you and then calls you by your mom's name. You correct her and she shakes her head and then says, "Oh that's right. Hello sweetie. Come give me a hug." Grandma continues to forget things throughout the day, and you wonder what is going on with her.

MEMORY LOSS

People talk about memory as though it is one thing, but memory is not one unitary thing. We talked about the different kinds of memory in Chapter 9. Although someone can suffer from a complete memory loss, where they don't know who they are, this is quite rare. More often, people suffer from a partial memory loss, where they know who they are but can't remember particular details of their life or can't form any new explicit memories. Memory loss can be permanent or temporary.

Amnesia is a form of memory loss. Hollywood loves a good amnesia story! Someone gets in some sort of accident, and then they suffer from some type of long-term amnesia where they can no longer remember who they are until the very end of the movie when everything turns out ok. Or someone has some type of short-term amnesia where they can remember everything that happens in a day, but once they go to sleep, it's like their memory has been reset, and they have forgotten everything all over again. While these types of storylines might make for a good movie, there's not a lot of science to back up what happens in many of these memory loss movies. Amnesia is a real thing, but it is not quite as simple as it is portrayed in the movies.

| Retrograde amnesia | Present | anterograde amnesia |

Amnesia can happen as a result of brain damage from a stroke, a car accident, concussion, or traumatic experience. People who cannot remember things that happened before their injury are said to suffer from **retrograde amnesia**, while those who cannot form new memories after their injury happens suffer from **anterograde amnesia**. Let's pretend that someone you know was in a serious car accident. If they woke up in the hospital and they couldn't remember what had happened in the hours leading up to the car accident, they would be suffering from retrograde amnesia. If they couldn't remember what happened after their car accident, even if you saw the person conscious and talking, they would be suffering from anterograde amnesia. While some people can recover certain memories with time and therapy, others never fully recover.

DEMENTIA

You might hear people say that someone is "suffering from **dementia**" and think it just means that they are losing their memory. But dementia is not any one specific disorder. Instead, it refers to a group of symptoms that not only involves memory loss, but also disordered thinking, poor problem-solving and communication skills, and deteriorating social skills that interfere with a person's everyday life.

Dementia is caused by damage to or loss of neurons and their connections in the brain. There are many disorders which may cause dementia. The exact symptoms that characterize a person's dementia depend on where the brain damage occurs. This is because, as we've seen throughout this book, different brain areas are involved in different types of behavioral functions. For example, if the frontal lobe is affected, a person might start to undergo personality changes that lead them to do things they never would have done before. They may lash out at loved ones or say inappropriate things. This can be hurtful to loved ones, but it is important not to take it too personally. People suffering from dementia can get overwhelmed easily by their environment and may not be able to communicate their physical or mental distress. Their outbursts may actually be cries for help. You can help them by trying to figure out what triggered their reaction, whether it was too many people talking at once or a chair that was making them feel uncomfortable.

ALZHEIMER'S DISEASE

The most common cause of dementia is **Alzheimer's Disease**. Alzheimer's Disease is named after psychiatrist and neuropathologist, Dr. Alois Alzheimer. A psychiatrist is a doctor who diagnoses, prevents, studies, and treats mental disorders. A neuropathologist is a doctor who studies diseases of neural tissues. Given his medical training, Dr. Alzheimer was in a perfect position to study how a patient's unusual behavioral symptoms might relate to their brain anatomy.

In 1906, a patient of Dr. Alzheimer died. She suffered from an unusual mental illness that included memory loss, unpredictable behavior, and language problems. Along with some of his colleagues, Dr. Alzheimer examined her brain and noticed something odd—her brain contained clumps of neurons, and many neurons were dead or looked damaged by tangled fibers within the neuron itself.

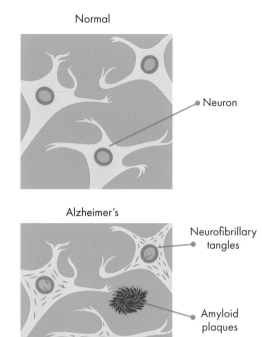

Normal

Neuron

Alzheimer's

Neurofibrillary tangles

Amyloid plaques

The abnormal clumps that are found in the brains of Alzheimer's Disease patients are what we now call **amyloid plaques**. Amyloid is a general term for the bits of proteins that are normally made by the body. Usually, these amyloid bits are broken down and expelled from the body as waste. Sometimes, however, these bits of protein stick around in the areas between neurons and build up, sticking to one another until a clump is formed. These amyloid plaques cannot be broken down and eventually cause cell death.

The damaged neurons that are found in Alzheimer's Disease patients are due to **neurofibrillary tangles** inside the cell. Neurons contain these long tubes called microtubules that help transport nutrients and other substances to all parts of the cell. The microtubule structures are dependent on a group of tau proteins for their shape. In Alzheimer's Disease, the tau proteins are abnormal and result in the microtubule collapsing. This results in the cell's eventual death.

While there is no current treatment, scientists are working hard to find ways to slow down or stop diseases like Alzheimer's Disease. Among the approaches they are studying, scientists are focusing on ways to break up the plaques and to stop the tau proteins from tangling. While far from perfect, there are some treatments that are being used to slow down the progression of some of the memory loss symptoms. In the meantime, if someone you know has dementia, whether it is from Alzheimer's Disease or some other disorder, the best thing you can do is to spend time with them and make memories that will last you a lifetime.

PICK YOUR BRAIN

As we age, our memories are not quite as sharp as they are when we are younger, but not all memory loss is an indication that someone has dementia.

A man named Clive Wearing has one of the most severe cases of amnesia documented. He suffers from both anterograde and retrograde amnesia and so cannot form any new memories, nor can he recall much from his past. He can only recall things that have happened in the last 30 seconds or so, and continually believes he has awoken from a comatose state, despite the fact that he has not.

MEET DR. BRAIN

Dr. Melonie Sexton is a cognitive neuroscientist who has conducted research on memory and attention. As the director of undergraduate research at her college, she helps students find their way into a lab. Recently, Dr. Sexton has begun working in the community with her students to improve literacy education and to increase access to educational resources.

Should we be worried when we forget things?

Believe it or not, forgetting some things can be helpful! My research has focused on the importance of forgetting. In fact, if we remembered everything, our brains would not be able to focus on the important things. We found that when people were so overloaded with remembering a bunch of things, they could not focus on what was important, and so they performed worse on memory and attention tasks. I have since taken my research and have used it to come up with "Tricks and Tips from Cognitive Psychology" to help my students learn how to study by showing them ways to focus on the important aspects of their study material.

How can you use psychology to help your community?

I believe in service-learning programs that apply what is being learned in the classroom to the community. I believe that it is a privilege to be educated and that we have a duty to use it for good. There are lots of ways that psychologists can help their communities through mental health education, police training programs, and school interventions. I use service learning in my college classes. I believe that service-learning programs are great for the college students, as well as for the community, who often don't realize that they can leave their footprints in the world, even if they are young.

Tell us about your service-learning projects.

My students and I work in our local elementary school to help increase literacy and improve access to educational resources. We want to show the students that education matters. We try to reinforce what they hear their parents and teachers tell them about the importance of a good education. We have completed several projects, including setting up three different Little Free Libraries in a homeless shelter, an elementary school, and a community center. A Little Free Library is something anyone can set up almost anywhere and is a place where people can take or leave books for people to read for free.

One day I mentioned in one of my classes how just 25% of children's books featured people of color as the main character. A couple of my students wanted to do something about this. We identified a local school with a high number of children of color in attendance. We paired up with a kindergarten class in that school, and my students and I created superhero stories with each child as the superhero. We interviewed the children to ask them which superheroes they liked and what superhero powers they would like to have. Then we wrote stories for each child. One thing we know about literacy is that people find books more interesting—and are therefore more likely to read them—if they can relate to the characters.

Dr. Melonie Sexton is a professor of psychology and director of undergraduate research at Valencia College in Orlando, Florida.

FINAL NOTES FOR YOUR NOGGIN

Memory loss can be either a permanent or temporary condition. Amnesia is a type of memory loss due to brain damage. Retrograde amnesia occurs when people can't remember things that happened before the brain damage, while anterograde amnesia occurs when people can't remember things after the brain damage. Dementia refers to a group of symptoms that include memory loss, disordered thinking, communication issues, and a loss of normal social skills that leads to disruptions in people's everyday lives. Dementia is the result of neuron loss or brain damage. The exact symptoms a person has depend on where the brain damage occurred. The most common cause of dementia is Alzheimer's Disease. The brain of a person with Alzheimer's Disease has abnormal clusters of neurons called amyloid plaques and has neurons that are dead or damaged from neurofibrillary tangles inside the neuron. Currently, there is no cure for Alzheimer's Disease or other disorders characterized by dementia, but there is hope that a cure is not far away.

MAKING MEMORIES

Memories are an important part of anyone's life. Memories of the past make us who we are today. But sometimes it's hard for us to recall exactly what happened on any particular day. And as we get older, we tend to forget the details of events that happened a long time ago, unless the day in question was the kind of day that really stood out in our memory. These types of memories are called **flashbulb memories**. But even with these memories, we sometimes don't remember everything perfectly. Despite this, we are usually very confident in our memories of these events! In this Lobe Lab, you will pick a day that was memorable for at least two people and compare their memories. Some ideas of events include the day of your birth, the last big holiday you celebrated, or some big news event.

MATERIALS NEEDED:
- Two people
- Your Brain Journal

STEPS:

1 Think of a memorable day and ask two people to recall the day's events to you separately. Ask them to include as many details as they can remember—what date/day of the week it was, time, what the weather was like, who was there, what happened when, and any other details they can remember.

2 Make a chart in your Brain Journal with two columns and a row for each of the questions you ask. Record each person's memories in the chart.

3 Ask each person to rate how confident they are that they remembered each of the day's events accurately on a scale of one (not very confident) to five (extremely confident).

4 Compare and contrast the two people's accounts of the day. What parts did they remember the same way? What parts did they remember differently?

WHAT HAPPENED?

You most likely found that the two people remembered some parts of the days in the same way, while other parts were remembered differently. Point out the differences to each person and see if one of them wants to change their recollection. How confident were the people in their memory for an event that each recalled differently? Oftentimes, we choose to believe our own memories over someone else's memories of the same event. This belief in our own memories comes from the idea that we believe that memory is an exact replica of a day's events. However, all our memories are reconstructions of what we believed to have happened that day, not necessarily what really happened that day.

NOW WHAT?

Repeat the experiment but include yourself this time. Ask a friend about an event you both attended and write down as many memories of that day's event as you can remember in your Brain Journal. Rate how confident you are about your memory for each event. Focus on the memories that you are sure happened one way, but that your friend recalled as happening differently. How confident are you that your memory is more accurate than your friend's?

CHAPTER 22

WHY DO I HAVE TO WEAR A HELMET

WHEN PLAYING SPORTS?

FRIDAY NIGHT LIGHTS

It's Friday night so you know what that means! Football at the high school stadium where your brother is the star quarterback. You make your way up to the fifth row of the home team stands where you are meeting your friends. The game starts and the center throws the football through his legs to your brother. He jumps back a few steps before reaching his arm back and then rockets the ball to the other end of the field into the opening hands of the receiver. The receiver runs 20 yards for a touchdown and the crowd goes wild! You're on your feet cheering and clapping when, all of sudden, you notice it is starting to get quiet. You scan the field and see your brother laying on the ground and not moving at all. The game stops as the coach and medics rush onto the field to check on him. After a few minutes, the medics help your brother up and he limps off the field to the sound of applause. You find out later that he got a concussion and you wonder what that means.

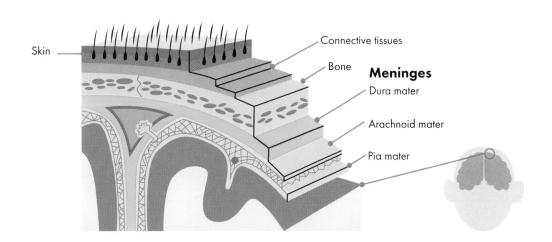

Skin • Connective tissues

Bone

Meninges

Dura mater

Arachnoid mater

Pia mater

Your brain is very delicate and has the consistency of jello. Think of carrying a bowl full of jello—it jiggles back and forth against the bowl whenever you move it. Your brain does the same thing. As you are walking and breathing, your brain moves ever so slightly in your skull. This is why your brain is protected by a thick skull (bone) and a thin layer of membranes (**meninges**) and fluid (blood and cerebral spinal fluid, layered between the meninges) that cushion your brain as it sloshes about. The meninges cover your entire brain and spinal cord. There are three layers of the meninges. The **pia mater** lies closest to the brain and is Latin for "tender mother." The middle layer is called the **arachnoid mater** and the layer closest to the skull is called the **dura mater**. Medical students remember the order of the layers using the acronym PAD (Chapter 4), which is appropriate because the meninges are used to *pad* the brain.

Most of the time, your skull, PAD, and fluid do a great job of protecting your fragile jello-brain. But if you get hit hard on the head, your brain will strike one side of your head with great force, causing it to bounce back on the other side of your head. It's like a tennis match in there with your brain as the ball. When the brain hits the inner walls of your skull, it can cause injury to your brain tissue, and can even cause bleeding of the brain. This type of traumatic brain injury (TBI) is called a **concussion** and it can have both immediate and long-term effects. People can sustain a concussion any time they get hit in the head, including when playing contact sports like football or soccer, being in a car accident, or even falling down and hitting their head on the ground.

BEHAVIORAL CHANGES FOLLOWING A CONCUSSION

As we will learn in the upcoming chapters, bleeding in the brain can be life-threatening, so someone who has been hit in the head should seek treatment as soon as possible. Some of the symptoms of a concussion are temporary, while others never go away (although they may get less severe with time and treatment). It's common for people to get a really bad headache and not remember what happened just before or after their accident. They might also feel confused or tired and may have a hard time concentrating or talking. Some people may even "see stars" or pass out. However, it's important to know that some of these symptoms might not even show themselves until hours or days after the person got the concussion.

It's common for kids (and adults) to get hurt while playing sports or while goofing around. Most of the time, all that happens is a scraped knee or banged up elbow, but if someone gets hit hard in the head, that's another story. If a person gets concussed, it's extremely important that they get checked out by a medical professional. Not getting help right away can delay a person's recovery or can lead to a second, worse, concussion. This is why it is recommended that athletes who get a concussion while playing their sport do not return to play the day they get hurt and that they get clearance from a medical professional trained in concussion evaluation before they return to the game.

PROTECTING YOUR BRAIN

Have you ever done the egg drop science project where you have to drop a raw egg from up high onto a solid surface with the goal of protecting the egg? Think of the egg yolk as your brain and the shell as your skull. Most people focus on protecting the egg by cushioning it with something soft like tissue paper or cotton. Sometimes they are successful, and the egg doesn't break. But if the egg does break, then the yolk gets destroyed. Athletes wear helmets to protect their head from being crushed, bruised, or scraped. You have to protect your skull so that it can help protect your brain. Many people are working on creating helmets that better protect the brain and head, but the existing helmets can only protect people if they wear them. So, it's always wise to wear a helmet when playing a sport and riding your bike. You only have one head with one brain, so you need to do everything you can to keep it healthy.

PICK YOUR BRAIN

The King-Devick test (K-D test) is a quick, two-minute test that athletes can take on the sidelines after getting hit hard in the head to check for concussions. Before a game, the athlete reads aloud single digit numbers that are printed on three index cards. If the athlete gets a blow to the head in the game, they read the numbers again on the sideline. If they are more than five seconds slower reading the numbers the second time, it can indicate a possible concussion.

PICK YOUR BRAIN

The Centers for Disease Control and Prevention is a United States federal agency whose main goal is to protect public health and safety by controlling and preventing diseases, injuries, and disabilities. It has developed a "Heads Up" Four Step Action Plan to protect athletes who are suspected of suffering a concussion while playing a sport. They advise that coaches should:

1. Take the athlete out of play immediately.

2. Make sure the athlete is evaluated by a medical professional and not try to assess the seriousness of the injury themselves.

3. Tell the athlete's parents about the suspected concussion and give them a fact sheet on concussions.

4. Keep the athlete out of play on the day of injury. Do not let the athlete return to play until they are cleared to play by a medical professional who is trained in evaluating concussions.

FINAL NOTES FOR YOUR NOGGIN

Your brain is protected by a thick skull and thin layer of membranes and fluid. Most of the time, this padding is enough to make sure that your brain does not sustain any injuries as it moves gently within your head. If your head gets hit hard though, the brain will hit the sides of your skull with great force, possibly causing a concussion. People who get concussions may suffer from headaches, a temporary loss of consciousness, memory and speech problems, and motor dysfunction. Athletes who get concussed in a game should not return to play until they are evaluated by a medical professional. Wearing a helmet helps protect your brain by protecting your head.

EGG HEAD

You can't just go out and get a concussion in order to study it. So, we'll do the next best thing in this lobe lab. We're going to do a version of the egg drop experiment, but instead of dropping it, we're going to shake the egg to mimic what might happen when someone gets hit in the head hard. The egg (shell and yolk) will represent your brain and the containers will represent your skull. Your observations will give you a good idea about what happens to your brain when you get hit in the head.

MATERIALS NEEDED:
- 2 raw eggs
- 2 plastic containers with lids
- Some water
- Paper towels (for clean up)
- Your Brain Journal

STEPS:

1 Before you do each of the following steps, take a moment to think about what will happen to the egg in each case. You could write your guess in your Brain Journal.

2 Take the first egg, place it in a covered plastic container, and shake it. Note what happens to the egg. Let's call this the Air Egg. Draw or write what happens to the Air Egg in your Brain Journal.

3 Take the second egg, place it in a covered plastic container with some water, and shake it. Note what happens to the egg. Let's call this the Water Egg. Draw or write what happens to the Water Egg in your Brain Journal.

WHAT HAPPENED?

Did you notice a difference between what happened to the two eggs? The Air Egg probably cracked up pretty good with the yolk going all over the place. This is what would happen to your brain if you didn't have your PAD and fluid cushioning your brain inside your skull. The Water Egg probably did not crack. This is because the water in the container helped to slow down the egg movement so that it did not hit the side of the container so hard. This is what happens in your head as you're walking along, bobbing your head, listening to music. Your brain is moving ever so slightly in your head, but your PAD and fluid protect it so that it doesn't hit the side of your skull and damage the brain tissue.

NOW WHAT?

Try shaking the Water Egg in its container hard enough that it cracks. You'll really have to shake it hard or smash it on a table. When (if) you finally crack the egg, note what the yolk looks like. The yolk is probably still in one piece, but it might be seeping out of the crack in the shell. This is what happens when you get a concussion. Sometimes, blows to the head are just too hard and your PAD and fluid can't protect your brain and it gets damaged. This demonstrates how much better your brain is with your PAD and fluid cushioning it.

CHAPTER 23

WHY DO I NEED PHYSICAL THERAPY?

SCOOOOOOORE!

It's the final game of the season and the Piranhas are doing their best to keep their hopes of the playoffs alive. You see your sister in the goal, and you can't believe how well she is doing! She's already made a couple incredible saves, diving to the left to stop a well-placed kick to the corner of the goal and catching another one as it approached the top of the net. You just hope her team can keep up the momentum and finish the game without another goal from the other team, which would cause a tie. It's the last two minutes of the second half and the opposing team is marching the soccer ball down the field toward your sister and the goal. The Piranha's defense is doing its best to slow their progress, but they just keep passing and dribbling the ball until, before you know it, they are in the goal box with only the goalie and one remaining defender between them and the goal. Your sister has a split-second decision to make—does she dive left or right? She dives right, *right* into the oncoming ball! She saved yet another goal! But instead of cheering, there seems to be chaos on the field. Your sister is on the ground clutching the ball in one hand and her head in the other. The remaining defender accidentally kicked her in the head while she was going for the save. You later learn that she sustained a concussion and now has to go to physical therapy. You wonder how that is going to help her.

PHYSICAL THERAPY

After suffering from a traumatic event, like a kick to the head, a car crash, or a stroke, people can lose the ability to do everyday things like walking, talking, or moving normally. Physical therapy helps people regain these abilities by retraining the brain and body to work together. Physical therapists work with people to gradually regain strength, motion, and activity after an injury. They first assess the person and their injuries to identify the weaknesses in the **biomechanics** of the body. For example, can the patients move their arms forward, but not backwards? Can they walk or balance or one foot? Can they follow a pen with their eyes? Once their weaknesses are identified, the physical therapist will design a treatment plan to get the patients "back to normal" as much as possible. Some people can be helped after one or two sessions with a physical therapist; others need several weeks of therapy. It all depends on the extent of a person's injuries.

Treatment plans involve specific exercises that target a person's weak areas in order to strengthen them. Although you may be able to look up how to do certain treatment exercises on the internet, treatment plans are specific to each individual and the specific injuries. In addition, physical therapists are highly trained individuals who are knowledgeable about the body's limitations and monitor the patient's heart rate and exertion levels so the person does not engage in activities that may make the injury worse.

Physical therapy requires a lot of hard work, time, and often frustration on the part of the patient. Imagine having to relearn how to walk, for example. It must be extremely challenging for someone to have to learn how to redo something that we all take for granted. Knowing how to walk is part of our procedural memory (Chapter 9). Procedural memory is part of long-term memory that is responsible for knowing how to do things—your motor skills. How you perform these motor skills is difficult to explain to others, which makes them hard to teach. Teaching someone how to walk again is just as hard as teaching someone how to ride a bike. There are a lot of tiny movements you make to adjust your **gait** that make it possible for you to remain upright and moving that you are not even aware of making. The goal of physical therapy is to make these adjustments become automatic again so that you don't have to think deliberately about taking the steps. Over time, and through experience, these movements will again become automatic.

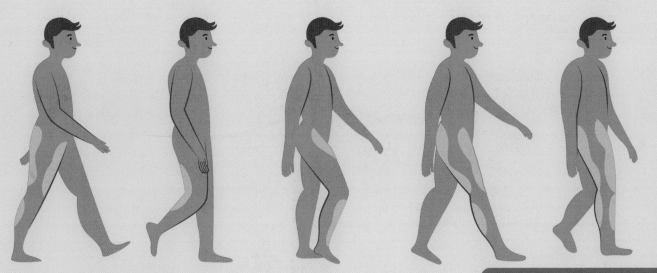

PHYSICAL THERAPY TREATMENTS FOR MOVEMENT PROBLEMS

Concussions happen when the brain is hit violently against the inside of the skull and results in tissue damage (Chapter 22). Concussions affect the body's homeostasis through the sympathetic and parasympathetic nervous systems (Chapter 19). In particular, blood flow is disrupted throughout the body, leading to changes in the way that nerves and muscles interact. Many treatments involve stimulating blood flow to get the nerves and muscles to work together again. **Ultrasound technologies** work by sending high-frequency vibrations into the deep tissues to get blood flowing again. **Electrical stimulation** works by stimulating nerves with the goal of altering the neural connections, leading to better muscle movements and stimulating blood flow.

Other treatments involve physical exercises that include the stretching and/or strengthening of certain muscle groups. Stretching helps to loosen up muscles. Strengthening helps improve one's range of motion and endurance. Exercises that work on your core muscles focus on those in your back, abdomen, and pelvis. These are the muscles that help you walk, sit up, and even go to the bathroom.

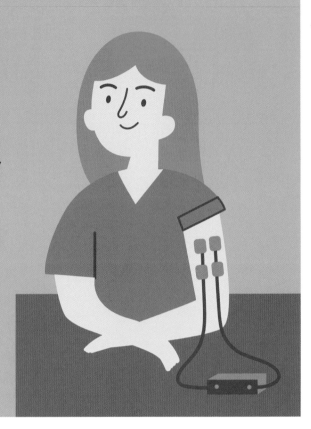

PHYSICAL THERAPY TREATMENTS FOR VESTIBULAR PROBLEMS

Many people who suffer from a concussion develop problems linked to their vestibular system, including feeling dizzy and not being able to walk or see straight because their balance is off. Remember that your vestibular system is responsible for telling you which way is up and whether you or something else is moving (Chapter 17). **Vestibular rehabilitation** is a class of physical therapy treatments that helps people with these types of vestibular problems. Two types of vestibular rehabilitation treatments include **gaze stabilization** treatments that help people to learn how to focus their eyes and **vestibular compensation therapy** that helps people to rely less on their vestibular system.

Under normal functioning, whenever you move your head to look at something, there is a coordinated movement of your eyes that allows you to keep your focus on the object you are looking at. Try this: look straight ahead at an object and, keeping your eyes on that object, move your head from side to side. In order for you to keep that object in focus, your eyes move in the equal and opposite direction of your head. To help a person see straight, gaze stabilization treatment teaches people how to keep their eyes focused even when they move their head.

Also, under normal functioning, as you are walking, your vestibular system is working with your visual system and your **proprioceptive system** to keep you upright and moving. Your proprioceptive system is the one that lets you know where your body is in space based on feedback from the muscles holding your body in place. For example, while standing up, close your eyes and slowly lift your leg out in front of you. You can probably sense where it is in the space in front of you and when you open your eyes, you probably know right where it is. This is because your vestibular system is letting you know how your balance is changing, your proprioceptive system is telling you where your leg is in space, and if you don't open your eyes within 20 seconds or so, you might fall down because your visual system needs to be able to give feedback to your nervous system about the location of your body and leg.

Vestibular compensation therapy is useful for people who have damage to the inner ear that results in the vestibular system giving conflicting information compared to the proprioceptive and visual systems. Remember when there is a mismatch between different sources of information, you can feel dizzy and even nauseous. The goal of vestibular compensation therapy is to have people rely more on their proprioceptive and visual system signals and to ignore their vestibular system signals. This kind of training takes a lot of time and experience to master.

Acute cerebellar ataxia is a disorder that can occur if the cerebellum is damaged. This can happen as a result of a stroke or a concussion. *Ataxia* means a person has problems with fine motor control. People with acute cerebellar ataxia have a hard time moving their arms, legs, and eyes, leading to problems like uncontrollable tremors occurring in their arms and legs, involuntary shaking of the head, and the inability to see straight or even walk.

MEET DR. BRAIN

Dr. Katie Dabrowski loves what she does! She is an athlete who took her passion for sports to the next level by becoming a physical therapist. Based on her past neuroscience research and her passion for endurance sports, she truly believes that movement is medicine. We asked her about physical therapy and how it works.

Why types of ailments do you treat in your practice?

I treat a variety of things which makes my job so fun! One group of people I see are active individuals, including high school athletes and weekend warriors. With this group, I focus on injury rehabilitation and prevention, as well as how they can optimize their athletic performance. A second group of people I see are those who are suffering from various types of orthopedic injuries like torn ACLs and muscle tears and strains. I see these people either after they have an operation or before they go into one to help them prepare for the surgical intervention. Finally, I meet with people who may have a neurological disorder such as those with spinal cord injuries, traumatic brain injuries (TBIs), or vestibular disorders. No matter which group people fall into, I talk with each patient one on one to figure out what their goals are and how I can help them keep doing what they love. My goal is always to figure out what is limiting my patient from being the best version of themselves and how I can help them achieve their goals.

What is the difference between injury rehabilitation and prevention?

Injury rehabilitation happens when someone needs to re-learn how to move after an injury. My patients take an active role in their therapy. I teach the patient exercises that they can do on their own to get better. Because of my background in neuroscience, I know that doing the movements is helping to rewire the patient's brain so that they can re-learn to do these movements on their own.

As far as prevention, I'd really like to get rid of the notion that a person should only see a PT when they have an injury. One of my jobs is to get people to move their body in the most efficient way possible for their lifestyle, whether that lifestyle involves playing baseball or gardening. I can assess a person's movement pattern to try to stop an injury before it happens. In fact, a lot of everyday aches and pains can be prevented just by learning to move the body in a different way.

Is this why you say movement is medicine?

Yes! When a person comes to see me, my prescription is always movement. Neuroscience and psychology research has shown that exercise—even walking—can help with not only our physical selves, but our mental and cognitive selves as well.

What is a typical session like with a patient?

It's really hard to come up with what a typical session is like because they are all so different. I tailor each of my sessions to each patient based on their reason for coming to see me. I have them perform certain exercises based on their current abilities and their end goals and make sure they do not injure themselves further in the process of healing.

What advice do you have for injury prevention?

You need to make sure you get a really good warm up and cool down before and after practice so that you can stretch the muscle groups that you'll be using. You also need to make sure that you're hydrating, eating well, and sleeping.

Katie Dabrowski, PT, DPT is a practicing physical therapist in Miami, Florida.

FINAL NOTES FOR YOUR NOGGIN

Physical therapy is needed when people are injured and can no longer perform normal everyday functions like walking, talking, or even sitting up. Physical therapists are highly trained individuals who assess people after they have been injured and help design and implement an individualized treatment plan for their patients. Physical therapy requires a lot of time and patience and its goal is to get the person back to their normal level of functioning as closely as they can. Physical therapy treatments include those that stretch and strengthen muscle groups and those that help a person relearn their behavior and brain to accomplish everyday tasks.

CHAPTER 24

WHAT DOES IT MEAN WHEN SOMEONE HAS A STROKE?

WHAT IS A STROKE?

You've had a long day, including a math test at school, a rigorous soccer practice after school, and when you got home, you still had a list of chores to complete. You finally have the chance to sit down and mellow out in front of the TV for a bit before bedtime. You turn to your latest TV show obsession, a medical drama. The usual emergencies start flooding into the ER over the opening credits when you notice that your favorite actor is guest starring in this episode as Mr. Bourne, who is being brought into the ER by his distraught daughter while she screams for help saying "he's having a stroke," and something about a "*fast* checklist." As you settle in for what appears to be an exciting hour of TV, you wonder what exactly a stroke is, and what that has to do with a "fast checklist." Let's explore.

CEREBRAL BLOOD FLOW

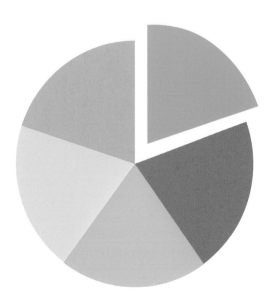

Your brain, like the rest of your body, needs blood to survive. This is because blood carries nutrients and oxygen, which are critical for survival, to all your organs. While cells in some organs, like your heart and kidneys, can survive for a bit of time without any or little oxygen, neurons are quite sensitive to a drop in oxygen levels. In fact, some neurons will die in a matter of minutes without oxygen. To safeguard against this happening, your brain gets a large proportion of your blood supply—about 15-20% of your total blood supply—even though the brain only makes up 3% of your body weight. That would be like your tiny 15-pound baby brother eating one slice of a five-sliced apple pie while the rest of your family, that together weighs 500 pounds, eats the rest!

Because blood flow is so important for your brain health, it is important that your brain's circulatory system is strong. And strong it is! At the base of the brain, there is a system of blood vessels that form a ring called the **circle of Willis**. Think of the circle of Willis like a merry-go-round. Blood enters the circle of Willis through one of many **arteries**, and the circle of Willis then distributes the life-preserving blood to the brain from here. The circle of Willis acts as a safety valve to ensure that blood still gets to the brain even if one of the arteries that supplies blood to the brain becomes blocked.

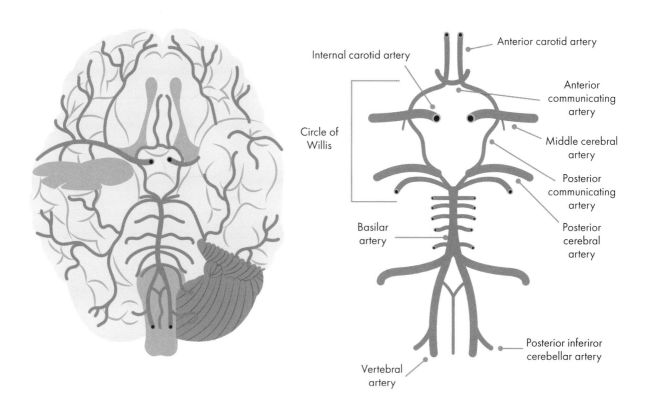

Blood flows to the brain through two pairs of arteries. Arteries are blood vessels that carry blood away from the heart to other parts of the body. The left and right **carotid arteries** are major blood vessels that travel up each side of the neck and supply the brain, neck, and face with blood. The word "carotid" comes from the Greek word *karotis*, which means "sleepy" because pressure placed on this neck artery can lead to sleepiness. The left and right **vertebral arteries** also travel up each side of the neck and join together in the skull at the base of the brain to form the **basilar artery**. The basilar artery then joins with the carotid arteries to form the circle of Willis. It's like there are multiple lines of happy blood cells waiting to ride on the circle of Willis merry-go-round!

BRAIN ATTACK

A stroke can be thought of as a "brain attack." Although there are many different kinds of strokes, all strokes involve blood being cut off from the brain, leading to neural damage or cell death due to the lack of oxygen. Nearly 75% of all strokes occur to people who are over the age of 65. Anyone at any age can decrease their risk of getting a stroke by eating healthy, exercising, and going for regular doctor check ups. Keeping your heart healthy doesn't have to be terribly hard—going on a daily walk after snacking on an apple can help quite a bit!

Ischemic stroke

Hemorrhagic stroke

The most common type of stroke is caused by a **blood clot**. A blood clot is a clump of blood that has hardened in your blood vessel. Whenever you scrape your knee or get some other kind of cut, a blood clot forms and acts like a plug so that you stop bleeding. You'll probably end up getting a scab, but once you've healed, the blood clot dissolves away. If a person doesn't have a healthy circulatory (blood) system, the blood clot doesn't dissolve and, instead, can move to other areas of the body, like the brain. An **ischemic stroke** happens when a blood vessel that is supplying the brain with oxygen is blocked by a blood clot. Another type of stroke, a **hemorrhagic stroke**, happens when a weakened blood vessel bursts and bleeds into the brain tissue. Whether a vessel is blocked or burst, the effect is the same—there is less oxygen reaching the brain.

SIGNS OF A STROKE

Since brain cells cannot survive long without oxygen, it is very important that a person suffering from a stroke gets treatment fast. A stroke often starts with something weird happening. A person might suddenly not be able to move their arm or smile and their speech may be slurred or nonexistent. Mr. Bourne's daughter said her father was reaching for the newspaper when, all of a sudden, he couldn't move or speak. She suspected her father was having a stroke and she knew that strokes can lead to permanent damage if help is not sought out right away.

This is what Mr. Bourne's daughter was talking about when she mentioned a "fast checklist." F.A.S.T. is actually an acronym to help you remember what to do if you suspect someone is having a stroke (it's a mnemonic like we talked about in Chapter 4). The checklist was developed by a group of medical professionals in the UK in 1998 as a way of training people to quickly recognize signs of stroke.

Facial drooping: can the person smile with both sides of their face?

Arm weakness: can the person hold up both arms for 10 seconds?

Speech difficulties: can the person speak clearly, and do they understand what is being said to them?

Time: if the answer to any of the above questions are NO, then it's time to call 9-1-1. Don't call a family member or friend, just call 9-1-1 and tell them you think the person is having a stroke. Time is of the essence, and it is critical that the medical professionals work **FAST** to avoid long-term damage.

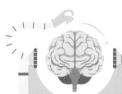

PICK YOUR BRAIN

Ways to reduce your chances of having a stroke:

Maintain a healthy weight:
Eat sensible portions of healthy foods.

Exercise more:
Go for a daily walk or join a sports team.

Keep making healthy choices throughout your life:
Don't drink alcohol or smoke now or when you are a teen or grown-up.

MEET DR. BRAIN

Scientists like Dr. Paige Brown Jarreau realize the importance of communicating science to the public. Dr. Jarreau started her career in biomedical engineering when she realized she missed writing and talking about science. She works at a company where she designs health-related apps and online educational platforms like LIFEApps.io that help people take control of their own health decisions. She is active on social media and has created a blog, *From The Lab Bench*, where she writes about all things science.

What made you decide to stop conducting your own research in favor of communicating all research?

When I was a kid, I wanted to be a writer. I loved to read, especially science fiction books, and one day I decided I wanted to be an author. Time went on and I was encouraged to study science because I really enjoyed doing kitchen experiments at home. I didn't realize that I could do science and write as a career. I started a degree in biomedical engineering, but I missed writing. I was doing science in the lab, but I wasn't seeing the whole story. It was then that I realized that I could combine my passion for both science and writing in a career in science communication.

What is science communication?

Many people think of science communication as only writing about science. This makes it seem as though science communication is very one-sided—I write it in my office, and someone reads it somewhere else. But science communication takes on so many forms! There are lots of people who create science videos like you see on YouTube, Hollywood often hires scientists to make sure that their movies and shows are scientifically accurate, and there is a great need for science-related apps that can help people live healthier lives.

Why is science communication important?

I believe scientists have a duty to communicate their scientific findings to the public. All people contribute to science by paying their taxes that help to pay for the research, as well as by supporting universities and companies where the science is conducted. Most people don't see the result of their support of scientific research, and I don't think that's right. Another reason is that our lives are getting more complicated in terms of available medical treatments, environmental concerns, and our own nutritional habits. Helping people understand basic science principles will help people make better, more informed decisions.

What advice do you have for someone interested in a career in science journalism?

The most important thing you can do is read a lot, especially science fiction and nonfiction. Reading helps to make you a good writer because you can see examples of good writing that may inspire you. You also need to practice writing often, even if you write in a journal that no one sees but you. Just write!

Dr. Paige Brown Jarreau is the Director of Science Communication and Social Media for LifeOmic, a company that works with healthcare and biotech organizations to facilitate research and help doctors deliver better patient outcomes. Visit her on Twitter @fromthelabbench.

FINAL NOTES FOR YOUR NOGGIN

FACE
Face look uneven?

ARM
One arm hanging down?

SPEECH
Slurred speech?

TIME
Call 9-1-1 NOW!

Strokes are caused by either a blockage in the blood vessels that supply the brain or by a blood vessel that bursts in the brain. This leads to a drop in the level of oxygen in the brain and, since neurons cannot survive long without oxygen, it is very important to get help fast. To help you remember what to do if you suspect someone is having a stroke, think of acting FAST: face, arms, speech, and time. If the person's face is dropping, they cannot hold up their arms, and/or their speech is disrupted, call 9-1-1 immediately. You might just save their life!

LOBE LAB

STROKE IMITATION

The most common form of stroke happens when a blood vessel is blocked, restricting or completely cutting off blood flow to the brain. In this lab, you'll see the effect that even a small blockage can have on blood flow.

MATERIALS NEEDED:

- Water
- Bowl
- $\frac{1}{4}$ measuring cup
- Straw
- Funnel (optional)
- Playdough
- Stopwatch
- Your Brain Journal

STEPS:

1 Do this activity with a friend, sibling, or parent. One of you should hold the stopwatch to keep track of how long it takes for the water to make it through the straw. The other should pour the water through the straw.

2 Slowly pour $\frac{1}{4}$ cup of water into the bowl through the straw. Be careful to get it all in the straw (a funnel might help). Note in your Brain Journal how long it takes for all the water to make it through the straw.

3 Plug one end of the straw with the playdough, but make sure water can still get through (only plug it up $\frac{1}{2}$ way). Once again, slowly pour $\frac{1}{4}$ cup of water into the bowl through the $\frac{1}{2}$ plugged up straw. Be careful, once again, to make sure all the water goes through the straw. Note in your Brain Journal how long it takes for all the water to make it through the straw. Is there a difference in the two times? If so, why do you think that is?

WHAT HAPPENED?

You should have noticed that the water took much longer to get through the plugged up straw. The playdough was acting like a blood clot, restricting the flow of water into the bowl. If the bowl was a brain and the straw was a blood vessel, you can see how the brain would be getting much less oxygen when there is a blockage in a blood vessel.

NOW WHAT?

There are at least two things that will affect how fast the water will go through the straw: the diameter of the straw and how much of the straw is blocked up. Try this experiment again with one of those big straws you get with your slushies. If you block it up half way with the playdough, does it take longer or shorter for the water to filter through the bigger straw? Can you plug up the big straw enough so that the time needed for water to go through the big straw is the same as the half plugged up small straw?

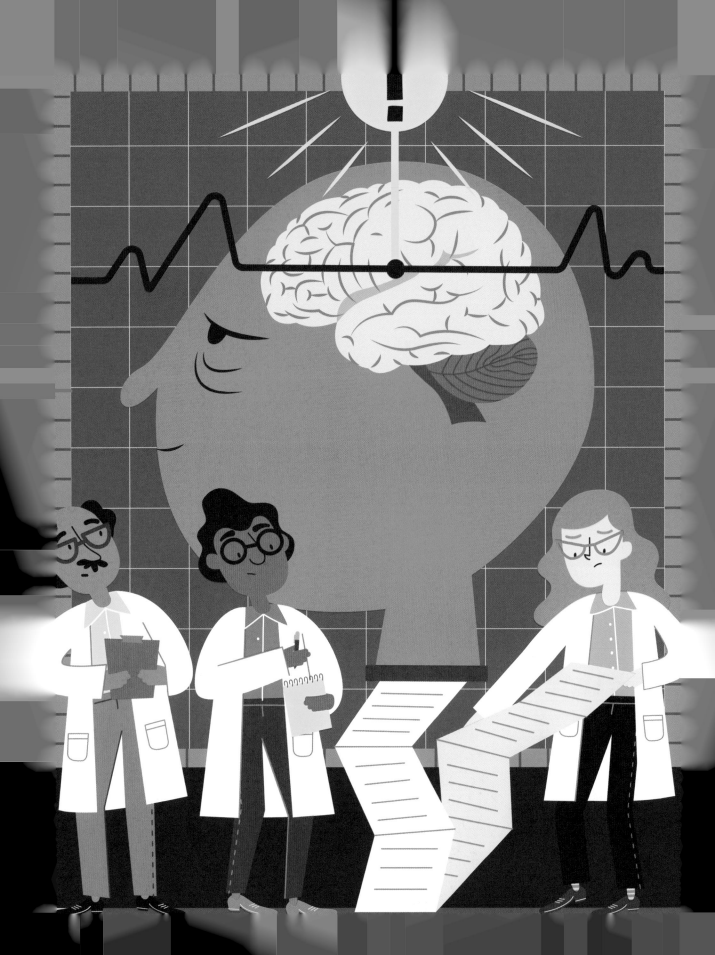

CHAPTER 25

WHY

CAN'T GRANDPA TALK NORMALLY

AFTER HE HAS HAD A STROKE?

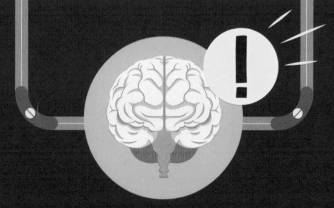

GRANDPA LOU'S HEALING

It's been pretty tough lately. Grandpa Lou had a stroke and is in recovery now. Actually, he's been in recovery for a while now and is doing much better. From what you can gather, Grandpa Lou was sitting at home listening to the radio with his wife when, all of sudden, he couldn't talk or move. His wife went over to him and noticed that half of his face was drooping. She called 9-1-1 and told them she thought her husband was having a stroke. The ambulance came and took him to the hospital where they performed all sorts of tests, including a brain scan. They gave him some medicine and kept him in the hospital for a few days. You visited him the day after the stroke, and he smiled at you, but something seemed off. Then, when you asked him how he was doing, his speech was slurred. You had to leave after that because the nurse came in and said he had to go to therapy. You visited him at his home a few weeks later, and Grandpa Lou gave you a big smile and told you to come over for a big hug. He looked and sounded much better, but you were still wondering what exactly happened?

BEHAVIORAL PROBLEMS AFTER A STROKE

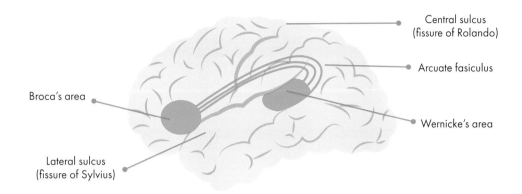

Central sulcus
(fissure of Rolando)

Arcuate fasiculus

Broca's area

Wernicke's area

Lateral sulcus
(fissure of Sylvius)

As you have read about in the previous chapters, the brain is a massive organ, but it is pretty well organized. There are specialized areas of the brain that make us move, think, feel, see, and all the other things we do. Neuroscientists talk about the structure and function of the brain— certain brain structures have certain behavioral functions. For example, the visual cortex structure allows you to see everything and your motor cortex structure allows you to move your body.

Given the brain's layout, the location of the stroke in the brain will determine what types of problems are seen after a stroke. One common problem stroke patients experience is difficulty initiating or understanding speech. The loss of the ability to speak or understand language due to brain injury is called **aphasia**. There are many areas of the brain that allow us to communicate. Two important areas are called Broca's area and Wernicke's area.

Broca's area is named after Paul Broca, a French doctor who described an area in the frontal lobe that we now know is important for speech production. In 1861, he heard of a man who had lost his ability to speak, even though he could understand when people spoke to him. After this man died, Broca examined his brain and discovered that the man had a **lesion** on a specific area on the left side of his brain. A lesion is an area where the brain has been damaged, kind of like a bruise on the brain. Broca treated several more people who suffered an inability to speak, despite being able to understand what was being said to them. All of these people had brain damage in a similar region of the brain—the region we now call Broca's area.

Another area important for speech is **Wernicke's area**, named after Carl Wernicke, a German doctor. Wernicke had several patients who had difficulty with their speech, but not in the same way that Broca's patients did. Wernicke noticed that some of his patients could speak just fine, but they did so in a rambling way, not making much sense. His patients had difficulty in understanding language and had brain damage in a region of the brain we now know as Wernicke's area.

Another common problem that stroke patients have is **paralysis**, or the inability to move the body. When a stroke happens, a person's body can suddenly feel like it has stopped working. You can see this in the face especially, since half the face seems to droop down. The person might also have a hard time walking or lifting their arms.

LATERALIZATION

Your brain is divided into two halves, or hemispheres. On each side of the brain, you have similar structures that perform similar functions. For example, the hippocampus on the right and left side of the brain both help you remember things like what you did on your last birthday. However, some functions are **lateralized**, meaning the structure on one side of the brain performs the function in a different way than the same structure on the other side of the brain. The movements of your body and language processing are good examples of lateralization. The structures of the brain that control these functions are on both sides of the brain but do slightly different things.

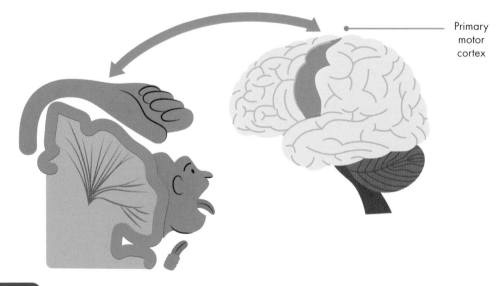

Primary motor cortex

In Chapter 11, we talked about how the somatosensory cortex contains a map of the body called the homunculus ("little man") that processes touch sensations from the opposite side of the body. Your motor cortex is set up in the same way. The left motor cortex map controls movement on the right side of the body, and the right motor cortex map controls movement on the left side of the body. Body movements are examples of lateralized functions since the two hemispheres control different body parts but do similar things, like move your arm or leg.

If a stroke happens in the left motor cortex, the person will have difficulty moving the right side of their body. Because the motor cortex has a somewhat orderly map of the body, the doctor can figure out *where* in the motor cortex the stroke happened by noticing which body part is affected. If a person can't move the left side of their face, then the doctor knows the stroke may have happened in the right motor cortex, near the bottom of the brain. If they can't move their right leg, then the doctor knows the stroke may have happened in the left motor cortex, near the top of the brain. A brain scan like a CT (Chapter 5) will be able to confirm exactly where the stroke happened in the brain.

Language processing is another good example of lateralization. In most people, the left hemisphere is where language is understood and produced, while the right hemisphere is where the emotional content of language is understood. If your friend said, "nice jacket" to you, your left hemisphere would understand what they said and allow you to respond, "thanks!" Your right hemisphere would help you decide whether their comment was sincere or sarcastic. Since language processing happens across many different areas of the brain, a stroke will most likely cause some difficulties in talking or understanding speech.

TREATMENT FOR STROKES

Treatment for strokes is dependent on many factors, including the type of stroke, the amount of damage it caused, and where the stroke happened. The good news is that our brains are "plastic." Not the hard plastic substances that are filling our landfills and oceans (recycle when you can!), but rather a property that the brain possesses (see Chapter 4). Plasticity in brain function means that the brain can re-route information around the injured areas of the brain to regain functions that were once lost.

Many forms of physical and speech therapy can help regain movement and language with a lot of hard work on the part of the patient. Some patients can learn to sit or stand or walk or speak again after several weeks of therapy, while other people will need therapy for months or even years to regain the functions they lost. The exact time frame varies from person to person. These therapies are essentially re-building the connections that were lost as a result of the stroke by retraining parts of the brain that are still healthy to take over for the parts of the brain that are injured. This process takes a lot of time, energy, and patience.

PICK YOUR BRAIN

Stroke patients who suffer from aphasia undergo intensive speech therapy that can last for quite some time. Aphasia does not affect the person's intelligence or the ability to learn and has no impact on how fast a person can recover their language skills.

Anomia is a form of aphasia in which a person can't name everyday objects like "telephone," "pencil," or "apple." It can be extremely frustrating since they know *what* they want to say, but they just can't seem to say it.

FINAL NOTES FOR YOUR NOGGIN

Grandpa Lou was lucky to get help soon after the stroke happened. Strokes cause muscle paralysis and language problems. Because of the way your brain is laid out, doctors can sometimes tell where in the brain the stroke happened based on what types of problems a person is having. A brain scan is used to confirm the doctor's suspicions. After a stroke, a person will go through intense therapy to retrain their brain so that they can learn how to move and talk normally again.

STROKE THERAPY

Many people recovering from a stroke find that they cannot move very well (or at all) and have poorer cognitive abilities, like thinking and paying attention after the stroke. They have to undergo serious physical therapy to help them learn to move, think, and talk again the way they did before the stroke. But you don't need a medical degree to help someone with their therapy. The key is to spend time with the stroke patient and get them thinking, moving, and talking. Here are some ideas that may help:

• **Play board games together.**
Games like Scrabble, Jenga, and card games help patients with their fine motor skills and help them to work on their problem-solving and attention skills.

• **Play mind games.**
Games that challenge the mind, like word searches, Sudoku, and putting together jigsaw puzzles help keep the mind active thinking about different possible solutions.

• **Do physical activities.**
Go for a walk, garden, dance, or do yoga together. These activities can bring a lot of joy and fun to the patient as they work on their balance and coordination and improve their cardiovascular fitness.

• **Do creative activities.**
Writing, cooking, painting, playing music, and any number of other creative expressions can help stimulate the brain by encouraging out-of-the-box thinking and allowing time for reflection as the person learns to live in their post-stroke life.

GLOSSARY

A

Action potential: the signal between neurons

Acute cerebellar ataxia: disorder that can occur if the cerebellum is damaged that causes people to have a hard time moving their body

Adapted (Adaptation): when we start to pay less attention to something because we have gotten used to it

Adrenaline: hormone in the body released by the sympathetic nervous system; helps with fight or flight response by increasing heart rate, enlarging blood vessels, and getting muscles ready to act

Alzheimer's Disease: a disease in which a person may suffer memory loss, unpredictable behavior, and language problems; neurons become damaged by amyloid plaques and neurofibrillary tangles

Amnesia: form of memory loss caused by brain damage

Amplitude: a factor in the loudness of sound; how big the sound wave is

Amygdala: almond-shaped brain structure located deep within the temporal lobes; involved in emotional processing

Amyloid plaques: formed by bits of protein not breaking down fully, sticking to the areas between neurons, and building into a clump; cannot be broken down and will cause cell death

Angular motion: motion of a body about a fixed point or fixed axis

Anomia: form of aphasia in which a person can't name everyday objects

Anosmia: the loss of sense of smell

Anterior cingulate cortex (ACC): area of the brain that processes pleasant information; located in the front of the brain, is involved in higher order cognitive functions like feeling empathy, decision-making, and impulse control; monitors our behavior and attentional focus

Anterograde amnesia: loss of the ability to form new memories after the event or injury that caused amnesia

Apex: part of the cochlea furthest from the oval window, responds best to low frequency sounds

Aphasia: loss of ability to speak or understand language due to brain injury

Arachnoid mater: the middle layer of membrane protecting the brain

Arcuate fasciculus: a bundle of nerve fibers that connect Broca's Area and Wernicke's Area

Arteries: blood vessels that carry blood away from the heart to other parts of the body

Astrocyte: a star-shaped glial cell that supports neural functioning and communicates with neurons by releasing a mineral wave

Attention: when we selectively concentrate on something at the expense of other things

Auditory canal: part of your outer ear that connects the outer ear to the eardrum; canal in which the sound waves travel to the eardrum

Auditory cortex: area of the brain that processes sound information; has tonotopic organization

Autobiographical memories: person-specific memories

Automatic behavior: a behavior that you automatically do without you thinking about it very much

Axon: a long nerve fiber that conducts action potentials

Babble: sounds that babies will make to practice using their vocal cords; for English-language learners this may include da, pa, and ma

Base: part of the cochlea near the oval window, responds best to high frequency sounds

Basilar artery: joins with the carotid arteries to form the circle of Willis

Bilingual: ability to speak two languages

Binocular field of vision: field of vision using both eyes; used for determining depth perception

Binocular rivalry: when the two eyes are sending the brain two different visual signals and the brain has to figure out which eye to pay attention to

Biomechanics: the study of the mechanics of the body and how it moves, such as the bending of an elbow or how someone can walk

Blood clot: a clump of blood that has hardened in your blood vessel; can cause strokes

Brain: organ that serves as the center of the nervous system in all vertebrate and most invertebrate animals

Brainstem: serves as a relay station between the brain and the spinal cord; sits beneath the cerebrum in front of the cerebellum; made up of the midbrain, pons, and medulla oblongata

Broca's Area: an area in the frontal lobe near the motor cortex that is important in speech production

Capsaicin: chemical in chili peppers that leads to a burning sensation and causes the release of Substance P

Carotid arteries: major blood vessels that travel up each side of the neck to supply the brain, neck, and face with blood

Caudal: back end of the brain

Central nervous system: system made up of the brain and the spinal cord

Central sulcus: longest and deepest dip in the brain, separates the frontal and parietal lobes

Cerebellum: part of the brain that controls balance and coordination; sits at the back of the head under the cerebrum

Cerebral ganglia: group of neurons that work together in one area of the brain

Cerebrum: largest part of the brain; controls memory, problem solving, thinking, and feeling

Circle of Willis: a series of blood vessels that acts as a safety valve to ensure that blood still gets to the brain even if one of the arteries that supply blood to the brain becomes blocked

Classical conditioning: type of learning where after repeatedly pairing two stimuli, the response to the second stimulus happens when only the first stimulus is presented

Cochlea: part of the ear that is made up of three fluid-filled canals surrounding a bony structure containing neurons that tell the central auditory system that a sound has happened; shaped like a snail shell

Code mixing: using words from two different languages in the same sentence

Conceptual vocabulary: total vocabulary across multiple languages; bilingual children may know fewer words in the common language, but their conceptual vocabulary would be the same as their friend who is monolingual

Concussion: traumatic brain injury in which your brain hits the inner walls of your skull, injuring your brain tissue and can even cause bleeding of the brain

Cone (cone cell): responsible for vision at higher light levels, involved in color vision and allows for fine spatial acuity

Congenital analgesia: rare condition in which a person cannot feel pain because the pain neurons in the spinal cord are unable to produce action potentials

Contralateral: having to do with the other side of the body

Coo: see babble

Coronal image: shows brain from front to back

Corpus callosum: bundle of nerves that connects the left and right hemispheres

Cortical magnification: describes the fact that in some cases, there are more neurons in your brain devoted to processing information from smaller areas of the body

Cranial nerve: 12 pairs of nerves that connect the brain to the body

CT scan: computerized tomography; combines a series of x-ray images taken from different angles around your body and uses computer processing to create cross-sectional images of the body

D

Dark adaptation: the adjustment of the eye to low light intensities involving reflex dilation of the pupil and activation of rod cells

Dementia: a group of symptoms that involves memory loss, disordered thinking, and deteriorating skills such as problem-solving, communication, and social skills

Dendrite: like branches of a tree off a neuron's soma; receives messages between neurons

Dermis: layer of skin that houses hair follicles and sweat glands

Diplopia: when you see two objects when there is only one because the brain cannot match the left and right images

Dopamine: a type of neurotransmitter that is responsible for signaling if something is rewarding or not

Dura mater: membrane layer that is closest to the skull

Eardrum: separates your outer and middle ears; a thin membrane that vibrates when sound waves hit it

Easy problems of consciousness: have to do with answering questions of how we pay attention to, discriminate between, and integrate information from the many different sources of information we get from our environment

Echolocation: sending out a sound and waiting for it to bounce back to the ear to determine location of object or animal

EEG: electroencephalogram; uses electrodes to record brain activity over time

Electrical stimulation: as related to physical therapy, electrical signals can stimulate nerves in the body to promote muscle movement and thus stimulate blood flow

Electrode: tiny wires that can record the electrical activity of the brain; placed on scalp or directly on or in the brain

Empirical: something that is observable and testable

Encephalization quotient: the animal's brain size relative to what they expect for that animal based on the complexity of the animal's sensory and motor behaviors

Epidermis: top layer of skin that you see, and the layer that creates your skin tone

Epilepsy: brain disorder that causes frequent abnormal brain activity, resulting in uncontrollable seizures

Epinephrine: see adrenaline

Episodic memory: a memory about a particular instance in your past, like your last birthday

Eustachian tube: part of the middle ear; tube that connects your middle ear to your throat to equalize the pressure in your middle ear to the air pressure outside your body

Excitatory neurotransmitter: make action potentials more likely

Explicit long-term memory: an explicit memory is something that you can describe where and how you learned it

Exposure therapies: a type of therapy that exposes someone to the thing they are scared of while a therapist talks them through it

fMRI: functional magnetic resonance imaging; MRI over time

Forebrain: largest part of the brain; includes the cerebrums

Free nerve endings: found in the skin, these feel sensations that are sent to your brain

Frequency: a factor in the pitch of sound; how fast the sound wave moves

Frontal lobe: helps make decisions; houses motor cortex

Functional brain scan: allows people to see how the brain responds over time, similar to a movie

G

Gait: the biomechanics of how a specific person moves to make their walk unique

Gate Control Theory of Pain Perception: scientific theory that explains how pain perception is controlled by both psychological and physical factors; posits a gate exists in the spinal cord that controls whether or not a pain signal is sent to the central nervous system

Gaze stabilization: a type of vestibular rehabilitation that teaches people how to keep their eyes focused when they move their head

Glial cells: a specialized cell that supports neurons by insulating them, cleaning up after them, and bringing the neurons oxygen and food

Gyrus (plural: gyri): mountains in the folds on the surface of the cerebrum

H

Hard problems of consciousness: have to do with answering questions of how neural activity gives rise to our experience of the world

Hemisphere: two halves of the brain, left side and right side, divided by the longitudinal fissure down the center of brain from front to back

Hemorrhagic stroke: when a weakened blood vessel bursts and bleeds into the brain tissue

Hindbrain: very back lower part of the brain rising up from the spinal cord; includes cerebellum and parts of the brainstem (including the medulla oblongata and the pons)

Hippocampus: seahorse-shaped brain structure that plays an important role in the limbic system and is involved with memory

Homeostasis: your body's state of equilibrium

Homunculus: distorted representation of the human body based on a neurological map of the brain processing motor functions or sensory functions for different parts of the body

Horizontal image: shows brain from top to bottom

Horizontal, superior, and posterior semicircular canals: see semicircular canals

Hunger: sensation felt when the body wants to eat

Hydrophobic: repellent to water

Hypodermis: bottom layer of skin that includes the fat and connective tissue

Hypothalamus: brain structure that helps control our feelings of hunger and thirst; works with the brain to maintain homeostasis

I

Implicit long-term memory: something that you cannot explain where you learned it or how to do it

Inferior prefrontal cortex: area of the brain involved in memory and impulse control

Inhibitory neurotransmitter: chemicals that make action potentials less likely

Introspection: when people describe their experiences as a way of measuring one's conscious life

Iris: colored portion of the eye; controls the diameter and size of the pupil

Ischemic stroke: happens when a blood vessel that is supplying the brain with oxygen is blocked by a blood clot

Knee-jerk reflex: involuntary kicking motion that happens after the knee muscle is stressed due to outside force

Lateralized: the tendency for one function to be controlled by only one side of the brain

Lens: helps focus the light on the back of the eye

Lesion: area where the brain has been damaged, kind of like a bruise on the brain

Limbic system: helps you control your emotions and remember things

Linear motion: motion that is backwards, forwards, or sideways and moves in a straight line

Lobe: four lobes on each side of the brain (frontal, parietal, occipital, temporal), all divided by sulci

Long-term memory: a memory that lasts a long time; scientists think that sleeping helps convert short-term memories into long-term memories

Macula: an area of the retina surrounding the fovea; gives us the ability to see and provides detailed vision

Mechanoreceptors: specialized receptors for touch; different mechanoreceptors respond to different kinds of touch information

Medulla oblongata: stem-like structure connecting the spinal cord with the brain; responsible for involuntary functions such as breathing and heart rate

Meninges: membranes that cover your entire brain and spinal cord

Microsleeps: brief episode of sleep lasting between a microsecond and ten seconds; usually occurs when someone is trying not to sleep

Midbrain: connects the hindbrain to the forebrain; contains structures that are important for moving your body and sensing your environment; part of your brainstem

Mind-body: the connection between the body and the mind; looking at how our mental states, events, and processes—like beliefs, actions, and thinking—are related to the physical states, events, and processes in our bodies

Mnemonic device: a memory technique that people use to help remember things

Monocular field of vision: field of vision using one eye

Motor cortex: housed in the frontal lobe; area of the brain that helps the body move

Motor neurons: sends information to your muscles to move

MRI: magnetic resonance imaging; uses electromagnetic pulses to get a very clear picture of the brain

Multilingual: ability to speak many languages, more than two

Myelin sheath: fatty material covering the axon; prevents the signal from leaking out of the axon

Neural plasticity: the brain's ability to change connections between neurons as we experience the world

Neurofibrillary tangles: due to abnormal accumulation of tau proteins within the neuron; will cause cell death

Neuron: a specialized cell responsible for thoughts and actions

Neurotransmitter: chemicals released between neurons, passing messages between neurons

Night Terror: extremely frightening nightmares

Nociceptors: a kind of free nerve ending that senses pain

Nucleus accumbens: a region in the basal forebrain that is an important part of the reward system, it releases the neurotransmitters dopamine and serotonin

Occipital lobe: lobe at the back of your brain; important for vision

Odorant: a substance that gives off a smell

Olfactory bulbs: part of the brain that processes smell information

Olfactory receptor cells: located in the roof of the mouth; responsible for the detection of smells

Operant conditioning: type of learning that leads to voluntary behaviors by rewarding desirable behaviors and punishing undesirable ones

Ophthalmoscope: tool doctors use to see in different structures of the eye; shines light into the pupil which reflects off the back of the eye

Optic disc: raised disk on the retina where the optic nerve leaves the eye, lacking visual receptors causing a blind spot

Organ of Corti: bony organ of the cochlea that converts physical sounds waves into action potentials

Ossicles: three bones in the middle ear, the hammer, anvil, and stirrup, that help transmit the sound waves from the outer ear to the inner ear; when sound waves hit the eardrum, the ossicles push on the oval window

Otolith organs: located in inner ear, contain small stones made of calcium carbonate crystals which help you know whether you are moving and the position of your body

Oval window: separates the middle and inner ears; membrane-covered opening that interacts with the ossicles when sound waves travel through the ear

Papillae: structures found on the tongue that contain taste buds

Paradoxical sleep: see Rapid Eye Movement (REM)

Paralysis: inability to move the body

Parasympathetic nervous system: your rest and digest system, directs your body's involuntary response in which it tries to calm itself and conserve energy

Parietal cortex: located on top of the brain; important for sensory information processing

Parietal lobe: behind the frontal lobe; houses sensory cortex

Peripheral nervous system: system made up of cranial nerves, spinal nerves, and peripheral nerves

Phenomenology: first-hand experience of the world

Phobia: an extreme fear of something, someone, or a situation

Photons: tiny particles that carry energy at different speeds, its movement makes light and what we see

Photoreceptor: cells in the retina that are sensitive to light; cones and rods

Pia mater: membrane that lies closest to the brain and is Latin for "tender mother"

Pinna: your outer ear; helps catch the sound waves

Polysomnography: measures heart rate, breathing, brain activity, and leg and eye movements; helps researchers determine which stage of sleep a person is in and how long they are in each stage

Pons: a brainstem structure that connects the medulla oblongata to the thalamus

Prefrontal cortex: located at the very front of the brain; important for many higher order cognitive functions including decision making, memory, attention, and personality expression

Procedural memory: action your body does without much thought; it's hard to explain how you do the activity after you learn to do it; activities include things like walking, running, or sitting

Proprioceptive system: the system in your body that lets you know where your body is in space based on feedback from the muscles holding your body in place; it works with your visual system to do things like walk

Proust Effect: emotional memories that are remembered after smelling something

Pupil: tiny hole in the eye that light enters; size determines how much or how little light to let in

Rapid Eye Movement (REM): a stage of sleep in which the eyes move rapidly under the eyelids; important for memory consolidation; also referred to as paradoxical sleep because your brain activity looks the same when you're asleep as it does when you're awake

Retina: structure of the eye that converts light energy into action potentials

Retrograde amnesia: loss of the ability to remember things that happened before a brain injury

Retronasal olfaction: when odorants move from your mouth up through your throat to the nose to be smelled; involved in the ability to perceive flavor dimensions of foods and drinks

Rod (rod cell): responsible for vision at low light levels

Rostral: front end of the brain

S

Sagittal image: shows brain from side to side

Semantic memory: memories of facts you have learned, like the 50 state capitols

Sensory neurons: send information about your senses back to the brain

Sensory specific satiety: decrease in pleasure one gets from eating the same type of food

Semicircular canals: three canals in the ear, named for their position within the ear; contains liquid that moves when you move and helps you orient yourself

Sequential bilinguals: people who learn a second language later in life

Short-term memory: a memory that lasts a short period of time, like what to pick up at the grocery store

Simultaneous bilinguals: people who grow up speaking two languages

Single-unit recording: measures the electrical activity of a single neuron; uses electrodes to record action potential of neurons

Sleep deprivation: not getting enough sleep; can be due to trouble falling asleep or staying asleep

Soma: neuron body that receives messages from dendrites; decides if message should be sent down the axon

Somatosensory cortex: part of the brain that processes all the sensations that you feel; located in the parietal cortex

Somnambulism: sleepwalking; occurs during stages 3 and 4 of sleep

Somniloquy: sleep-talking; occurs during stages 3 and 4 of sleep

Spinal cord: carries messages between the brain and the body

Stages of sleep: stages into which a person enters while sleeping; five stages

Stereopsis: the ability to see in depth because your brain gets two slightly offset images

Structural brain scan: allows people to see the structures of the brain, similar to a picture

Substance P: neurotransmitter involved in the perception of pain

SULCUS (PLURAL: SULCI): valleys in the folds on surface of the cerebrum

Sympathetic nervous system: your fight or flight response, directs your body's involuntary response to a dangerous or stressful situation

Synaptic terminal: located at end of axon; terminal where the outbound message between neurons is sent; releases neurotransmitters into space between neurons

T

Temporal lobe: underneath the other lobes; important for hearing and memory

Thermoreceptors: a kind of free nerve ending that senses temperature

Thirst: sensation felt when the body wants to drink

Tilt motion: motion that tells you which way is up or down

Tonotopic organization: a type of neural organization in which cells are located in an orderly fashion based on the different frequencies they respond to

Two-point threshold: the shortest distance at which two points on the body can be distinguished from one point

U

Ultrasound technologies: related to physical therapy, ultrasound technologies can send vibrations deep under the skin to get blood flowing

Unihemispheric sleep: type of sleep where one half of the brain sleeps and the other half stays awake to sense environment

Utricle and saccule: the two otolith organs

V

Ventral: the bottom of the brain

Ventral tegmental area: located in the midbrain, it is a group of neurons that release dopamine as part of the reward system

Vertebral arteries: travel up each side of the neck and join together in the skull at the base of the brain to form the basilar artery

Vertigo: when the vestibular system is not matching what your visual system is seeing; feeling like you are spinning or dizzy

Vestibular compensation therapy: therapy that allows people to rely more on their proprioceptive and visual system signals and ignore their vestibular system signals; for people with damage to their inner ear that caused their vestibular system to conflict with their proprioceptive and visual systems

Vestibular rehabilitation: a class of physical therapy treatments that help people who cannot see straight or walk without feeling dizzy; two types include gaze stabilization and vestibular compensation therapy

Vestibular system: the system that communicates to your brain which direction is up or down and your spatial orientation; helps you see clearly and stay upright

Visual field: how much a person can see without moving their head

W

Wernicke's Area: an area in the temporal lobe near the auditory cortex that is important in understanding speech

INDEX

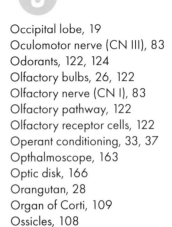

ABOUT THE AUTHOR

Leanne Boucher Gill, PhD, is a professor of psychology at Nova Southeastern University, where she received the Faculty Excellence in Teaching Award and was named the NSU STUEY Professor of the Year. She maintains an active research program studying how exercise affects the way we think. She lives in South Florida.

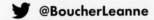 **@BoucherLeanne**

Magination Press is the children's book imprint of the American Psychological Association. Through APA's publications, the association shares with the world mental health expertise and psychological knowledge. Magination Press books reach young readers and their parents and caregivers to make navigating life's challenges a little easier. It's the combined power of psychology and literature that makes a Magination Press book special.

Visit maginationpress.org

    @MaginationPress